# PLANT NUTRIENT DISORDERS 2
Tropical Fruit and Nut Crops

# PLANT NUTRIENT DISORDERS 2
# Tropical Fruit and Nut Crops

R.G. Weir

G.C. Cresswell

Biological & Chemical Research Institute

NSW Agriculture

Inkata Press

Melbourne

INKATA PRESS
A DIVISION OF BUTTERWORTH-HEINEMANN

| | | |
|---|---|---|
| AUSTRALIA | BUTTERWORTH-HEINEMANN | 18 Salmon Street, Port Melbourne 3207 |
| UNITED KINGDOM | BUTTERWORTH-HEINEMANN LTD | Oxford |
| USA | BUTTERWORTH-HEINEMANN | Stoneham |

National Library of Australia Cataloguing-in-Publication entry

Weir, R. G.
  Tropical fruit and nut crops.
  Bibliography.
  ISBN 0 909605 90 4.
  ISBN 0 909605 88 2 (series).

  1. Tropical fruit – Nutrition. 2. Nutritionally induced diseases in plants. 3. Tropical fruit – Diseases and pests. 4. Nuts – Nutrition. 5. Nuts – Diseases and pests. I. Cresswell, G. C. (Geoffrey Charles), 1953- . II. Title. (Series: Plant nutrient disorders; 2).

634.0493

©1995 NSW Agriculture
Published by Reed International Books Australia Pty Limited trading as Inkata Press

This book is copyright. Except as permitted under the Copyright Act 1968 (Cth), no part of this publication may be reproduced by any process, electronic or otherwise, without the specific written permission of the copyright owner. Neither may information be stored electronically in any form whatsoever without such permission.

Enquiries should be addressed to the publishers.

Designed by John van Loon

Edited and produced by Patricia Sellar Editorial Services

Typeset in $10^{1}/_{2}$ pt Goudy Old Style by MACKENZIES, Melbourne

Printed through Bookbuilders

# CONTENTS

| | |
|---|---|
| Acknowledgments | vi |
| Preface | vii |
| | |
| 1 – PLANT NUTRIENT NEEDS | 1 |
| Causes of nutrient deficiency | 1 |
| Causes of nutrient toxicity | 3 |
| Nutrient imbalances | 4 |
| Crop sensitivity and tolerance | 4 |
| | |
| 2 – IDENTIFYING NUTRITIONAL PROBLEMS | 5 |
| Step 1 – Gathering the facts | 6 |
| Step 2 – Diagnosis from visible symptoms | 8 |
| Step 3 – Confirming the diagnosis | 11 |
| Step 4 – Correcting the problem | 12 |
| Step 5 – Following up | 12 |
| | |
| 3 – PLANT ANALYSIS | 13 |
| Sampling method | 14 |
| Making a two-sample comparison | 15 |
| Quick sap tests | 15 |
| Adjusting the fertiliser program | 19 |
| | |
| 4 – COMMON NUTRITIONAL PROBLEMS AND THEIR CORRECTION | 20 |
| Deficiencies | 22 |
|   Nitrogen | 22 |
|   Phosphorus | 28 |
|   Potassium | 30 |
|   Calcium | 38 |
|   Magnesium | 41 |
|   Sulphur | 46 |
|   Iron | 48 |
|   Manganese | 53 |
|   Zinc | 56 |
|   Copper | 62 |
|   Boron | 65 |
|   Molybdenum | 69 |
| Toxicities | 70 |
|   Sodium and chloride | 70 |
|   Boron | 73 |
|   Manganese | 75 |
|   Nitrogen (fertiliser burn) | 78 |
| Some non-nutritional symptoms | 79 |

| APPENDIX | 85 |
| --- | --- |
| Leaf sampling | 85 |
| Avocado | 88 |
| Banana | 89 |
| Coconut | 90 |
| Coffee | 91 |
| Custard Apple | 92 |
| Guava | 93 |
| Kiwifruit | 94 |
| Lychee | 95 |
| Macadamia | 96 |
| Mango | 97 |
| Passionfruit | 98 |
| Pawpaw | 99 |
| Persimmon | 100 |
| Pineapple | 101 |
| References | 103 |

## Acknowledgments

The authors wish to thank those people who assisted in providing information and photographs during the preparation of this book: Messrs Roger H. Broadley, Garth Sanewski and David Swete Kelly of the Department of Primary Industries Queensland; David Stevenson, R. Fitzell, D. Batten and G. Johns of NSW Agriculture; and Mr F.C. (Fred) Chalker (formerly NSW Agriculture). Several photographs were reproduced with permission from the Queensland Department of Primary Industries Information Series publications *Custard Apples – Cultivation and Protection* Q190031 and *Pineapple Pests and Disorders* Q192033. Thanks are also due to Mr L. Turton of NSW Agriculture for processing many of the photographs.

# PREFACE

*T*his manual is part of a series called Plant Nutrient Disorders that aims to help farmers, advisers and students to identify nutrient deficiencies and toxicities in a wide range of crops and pasture plants.

In all, there are five manuals dealing with the identification of nutritional disorders, and they are:
1. Temperate and Subtropical Fruit and Nut Crops
2. Tropical Fruit and Nut Crops
3. Vegetable Crops
4. Pastures and Field Crops
5. Ornamental Plants and Shrubs

Each book describes and shows, with the aid of coloured plates, typical symptoms of nutrient deficiencies and toxicities found in a range of common crops. The techniques for distinguishing symptoms caused by a nutrient deficiency or toxicity from some other problems, such as virus disease, herbicide damage or moisture stress, are explained and illustrated.

These methods evolved over thirty years of close collaboration between plant nutrition chemists, advisory horticulturists and agronomists from the New South Wales Department of Agriculture in their efforts to find solutions to many farming problems. This service ceased in 1988, but it is hoped that the skills developed for diagnosing nutrient disorders in plants will be passed on through each of these books.

Many of the coloured plates used in the manuals are of field specimens collected from affected crops where leaf analysis was used to identify the problem. Analytical records used to develop some of the leaf analysis standards in these books go back to 1958 and cover probably the widest range of crops available to date.

This series provides leaf analysis standards and sampling procedures for many common crops, as well as some poorly documented species. Where possible, sampling times are given as a growth stage or season, but when particular months are quoted they refer to Southern Hemisphere crops. The corresponding Northern Hemisphere times follow in brackets. The manuals also explain how the leaf analysis technique can be used, how to interpret analyses and how leaf analysis should not be used. This information should be helpful to anyone who wishes to build diagnostic and interpretive skills.

# CHAPTER 1
# PLANT NUTRIENT NEEDS

**S**ixteen elements are known to be essential for the normal growth of green plants. Three of the elements – carbon, hydrogen and oxygen – are obtained from the air and from water in the soil. The remainder – nitrogen, phosphorus, potassium, sulphur, calcium, magnesium, iron, manganese, copper, zinc, boron, molybdenum and chlorine – are obtained chiefly from the soil. Other elements, such as silicon and sodium, have been shown to improve the growth and health of certain plants in particular conditions, but field-grown tropical fruit crops rarely respond to these elements.

Plants need large amounts of nitrogen, phosphorus, potassium, sulphur, calcium and magnesium which are called 'major' or 'macro', nutrients. The remaining elements are needed in much smaller amounts, and are known as 'minor' or 'trace' elements.

All nutrients are normally present in the soil to some extent. They come from the parent rock, from decomposing organic matter and from fertilisers. Plants absorb these nutrients from the soil solution and transport them to leaves and other parts where they are used for growth and tissue maintenance. Each nutrient is involved in physiological processes essential for plants to function normally. Consequently, if there is too little or too much of any one nutrient, plant health suffers, leading to slow growth, low yields, or poor quality, and symptoms are exhibited on leaves or other tissues. There are two types of nutrient disorder:

- **Deficiencies**   where there is too little of an essential element to sustain optimum plant performance.
- **Toxicities**   where the supply of an element is more than the plant can tolerate.

Figure 1 shows the general relationship between the supply of an essential element and crop performance.

## Causes of nutrient deficiency

The two main reasons for a deficiency are that the amount of a nutrient in the soil is low or that it is not in a form available for plant uptake.

- **A low level of nutrient in the soil**

Some soils, especially those derived from sedimentary rocks like sandstones and shales, are naturally poor in nutrients. The most fertile

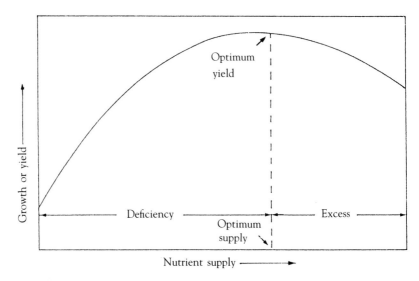

**Figure 1 Relationship between crop growth or yield and nutrient supply**

soils are formed from weathered igneous rocks rich in minerals, or from mineral-rich material deposited on river flood plains. Nutrients are lost from the soil through leaching and erosion, and become depleted especially in high rainfall areas. For many Australian soils these processes have been operating over a very long period, so the soils are naturally low in many nutrients. Some farming practices, especially those which leave the soil bare and vulnerable to erosion and leaching, can accelerate nutrient loss. Significant quantities of nutrients are often removed in crops, leading to further nutrient depletion of the soil unless they are regularly replenished by balanced fertilisers and manures.

- **The nutrient is not present in a readily available form**

Plants absorb nutrients primarily from the soil solution and so nutrient forms which are not very soluble, for example some mineral phosphates, are not readily available to the plant. The acidity or alkalinity of the soil (measured by its pH) affects nutrient availability. Deficiencies are more likely to develop in soils that are alkaline or very acid than in soil where the pH is neutral or only slightly acid. For example, zinc and iron deficiencies occur in sensitive crops planted in many of the naturally alkaline soils in inland New South Wales, while on many coastal acid soils molybdenum deficiency is a problem of some vegetable crops and pasture legumes.

Soil pH can change because of farming practices, for example, deficiencies of boron, iron, manganese and zinc can develop in soils made too alkaline by over-liming. If soils become too acidic, the availability of nutrients like calcium, magnesium, nitrogen, phosphorus and molybdenum is reduced. The long-term use of sulphate of ammonia is a farming practice which can cause excessive acidity, and lead to deficiencies of one or more of these nutrients.

# Causes of nutrient toxicity

Soil acidity and salinity are the two major causes of toxicities in crops and pastures. In acid soils, toxicities of manganese and aluminium often occur. Under saline conditions, chloride and sodium toxicities are common.

Many soils are naturally acid and may become even more acidic after several years of cultivation, irrigation and fertiliser use. The increased availability of elements such as manganese and aluminium at low pH can lead to toxicity. High manganese availability can also be a problem in waterlogged soils and in some soils affected by heat and drying during summer fallow. The solubility of aluminium increases in strongly acid soils causing severe stunting of the roots of sensitive plants.

Nitrogen is one of the most important of the nutrients normally applied to crops. However, most nitrogenous fertilisers – ammonium sulphate, urea, ammonium nitrate and mixed fertilisers containing ammonium salts – acidify the soil. Ammonium sulphate (sulphate of ammonia) is the most acidifying and should be avoided for most soils. Even the nitrogen fixed by legumes or from manures can contribute to soil acidification. This happens because some of the plant-available nitrogen is leached from the soil as soluble nitrate, a process which also removes calcium and magnesium.

Many soils have naturally high levels of salt, especially in arid and semi-arid areas, where rainfall is too low to leach it away. Salinity problems quickly develop in these areas if the native timber is cleared or irrigation is used without adequate drainage. These practices can raise the watertable bringing salts up into the root zone. Salts of chloride and sodium leached from soil by irrigation water are concentrated by evaporation along edges of furrows, in low areas or in seepage areas above impervious subsoil layers.

Some salt- or boron-affected soils have developed from rocks formed under marine seas or large inland lakes which became saline as they dried up. Irrigation water from bores, wells, dams or rivers, especially at times of low river flow, are other sources of chloride and sodium. Water to be used for irrigation should be tested for salinity. Although fertilisers are not a major factor in salinity, avoid muriate of potash (potassium chloride) and fertilisers containing it where salinity is a concern, for they may add to the salinity problem.

Toxic concentrations of fertiliser salts near the roots from the careless or excessive application of fertilisers can also cause injury. Germinating seedlings and young plants are especially sensitive and are commonly damaged when concentrated soluble fertilisers are banded too close to the roots or placed in the planting hole of young trees.

Under special circumstances, toxicities of some of the trace elements can occur. Boron, copper and zinc have narrow optimum ranges (between enough and too much) and should be used carefully. Sawdust derived from softwood timber treated with boron for pest control has caused toxicity when used as mulch around sensitive plants.

Heavy metal residues in mining and sewage wastes are other potential causes of toxicity.

Uneven application or use of high rates of trace elements can easily lead to toxicity. Special care must be taken when applying borax. Pockets of high boron concentration can be avoided by crushing all lumps and spreading the powder evenly over the area to be treated. Foliar sprays of trace elements can cause injury if product recommendations are not followed. Soluble copper salts can be very phytotoxic when applied as sprays and are often 'neutralised' with hydrated lime to avoid foliage burn.

## Nutrient imbalances

Plants need a proper balance between the various nutrients to grow well. Too much potassium may interfere with the uptake of magnesium or calcium, leading in some cases to deficiency of one or both of these elements. In the same way, an excess of phosphorus fertiliser can induce zinc or iron deficiency if these elements are in short supply.

## Crop sensitivity and tolerance

Plants differ in their sensitivity to specific nutritional problems. For example, boron deficiency is more common in pawpaw than in passionfruit or sugarcane. Macadamia, persimmon and passionfruit are more sensitive than bananas or lychee to high levels of manganese in the soil. While magnesium deficiency is rarely encountered in kiwifruit, it is common in banana, macadamia and custard apple trees growing in the same areas (see Table 4).

Differences in plant sensitivity to a deficiency arise because some species or varieties need more of a nutrient for healthy growth. Others are less efficient at obtaining the element from the soil or of using what they have absorbed. Crops also differ in their tolerance of toxicities. This is related to differences in their abilities to exclude elements from uptake at the root surface, or to keep absorbed toxic elements away from sensitive tissues or render them less biologically harmful. Both the rootstock and the scion can affect a crop's tolerance or sensitivity to specific toxicities or deficiencies.

# CHAPTER 2
# IDENTIFYING NUTRITIONAL PROBLEMS

Plants suffering from a nutrient deficiency or toxicity will often develop visual symptoms that are useful for identifying the cause of the disorder. These visible effects include stunting, abnormal growth, unusual colours, chlorotic patterns or burns in the leaves, or distortion of individual plant parts.

As some non-nutritional factors can produce similar symptoms (Table 1), careful observations are needed to ensure that the diagnosis is reliable.

In higher plants, nutrients move from the roots to other parts of the plant through a network of cells called the vascular system. These tissues are specialised in the transport of water, nutrients and metabolic products throughout the plant. The most obvious part of this vascular system are the veins in leaves. The arrangement of veins and the ease with which individual elements move within the plant (mobility) strongly influence the way symptoms develop. In manganese-deficient leaves, for example, leaf blade tissues close to the major veins are last to become chlorotic because they have first call on manganese available in the sap.

The close relationship between the symptom pattern and the arrangement of the veins is an important feature of nutrient-related problems and helps to distinguish them from most symptoms of non-nutritional causes. Symptoms caused by the latter usually show no relationship to vein pattern.

Table 1  Major causes of visual symptoms in plants

| Nutritional disorders | Other disorders |
| --- | --- |
| deficiencies<br>toxicities | infectious diseases – fungal, bacterial or viral<br>insect damage<br>physiological–environmental stresses<br>mechanical injury<br>chemical injury – pesticide, air pollution, spray burn |

> **Characteristics of nutritional symptoms on leaves**
> - Restricted initially to a single leaf-age class, that is, young, old or intermediate aged leaves.
> - Patterns are symmetrical and closely related to leaf venation.
> - Changes in leaf colour and tissue death develop gradually (rarely overnight).
> - Boundaries between green and chlorotic areas on a symptom leaf tend to be diffuse. Strong, definite boundaries are often produced by herbicides or viruses.
> - Leaf patterns are rarely blocky or angular. Such patterns can be caused by a pathogen or occasionally by nematodes.
> - Damage to the surface of a symptom leaf is unusual. Nutritional problems impair cell function and rarely cause mechanical disruption of the cuticle.
> - Symptoms develop first in tissues most distant from the major veins of the leaf, for example, the interveinal regions, tips and margins of the leaf blade.

After establishing that the symptoms are nutritional, and not due to pests, diseases or other causes, take the following steps to identify the disorder and, more importantly, find its underlying cause.

## Step 1 –
# Gathering the facts

Background information is needed to define the problem and to identify the most likely factors contributing to its development. Some causes can then be discounted, narrowing the scope of the investigation. Furthermore, this information can help establish the underlying cause of a problem, which is crucial when developing an appropriate corrective treatment.

**Field description**  Briefly describe the overall appearance of the crop, highlighting obvious abnormalities such as unusual leaf colouring, stunting or thinning in the crop. Relationships between the distribution of problem areas in a crop and geographical features, such as soil type, fence lines, crop rows or irrigation bays, may point to a soil or nutritional cause.

**Severity of problem**  Estimate the percentage of the crop currently affected and the expected reduction in yield. This information may be useful for deciding between possible causes. For example, a mild manganese deficiency is unlikely to be the primary cause if yield has been reduced by 50%.

**Crop**   Details of cultivar, rootstock, tree age and overall crop health are helpful in assessing the problem and essential for interpreting plant test results. Sensitivity to a particular nutrient deficiency or toxicity often differs with cultivar or rootstock.

**Crop developmental stage**   Note the stage of growth, for example, vegetative, flowering, or fruit filling. These details are needed for interpreting plant analysis results, but the growth stage when a symptom is first expressed can also give a significant clue to the cause of the disorder.

**Contributing factors**   Any change in management immediately before the symptoms developed should be noted. The time taken for the symptom to become fully expressed may also provide a clue to the cause. For example, symptoms which develop suddenly are usually caused by catastrophic events such as frost, wind, pests, and herbicides or chemical sprays.

**Cropping history**   Record previous crops grown on the site, their management and performance. Residual chemicals in the soil, such as herbicides, liming materials and fertilisers, can affect later crops.

**Soil type and depth**   Are the soils uniform in type and depth across the problem area? Is there a relationship between the distribution of affected plants and variation in soil properties? Are affected patches related to land levelling, former roadways, fence lines, irrigation banks, or log burning? Relationships between the occurrence of a disorder and soil type or previous land use often indicate soil chemical or physical problems. Examine the soil to the appropriate root depth in a problem patch for a possible hard pan, shallow rock or clay.

**Irrigation type/frequency**   Is irrigation practised and, if so, what type: overhead, furrow, trickle or capillary? Has the crop been moisture stressed? Symptoms resulting from severe water stress may be confused with nutritional disorders.

**Drainage**   Are the soils well drained? Waterlogging can lead to nutritional disorders including manganese toxicity, iron deficiency and nitrogen deficiency, or produce wilting, leaf fall, vein chlorosis and hormone-like symptoms of new growth.

**Weather conditions**   Were weather conditions leading up to the problem unusual, for example, periods of heavy rain, drought, high or low temperatures, or frost? Both current weather and that occurring some months earlier, may be significant.

**Fertiliser history**   Problems can sometimes be traced to a recent change in fertiliser practice, to a history of over- or under-use of fertiliser, or of an unbalanced program. Fertilisers and the rates in current use should be compared with district recommendations and with those used by other growers in the area.

**Paddock history**   Does the problem area coincide with a previously cropped or fertilised section of the block?

**Spray program**  What chemical sprays have been used on this crop – pesticides, nutrients or other? Pesticide and nutrient sprays can burn foliage, flowers and fruit, if not used according to the manufacturer's recommendation. Nutrient sprays and some fungicides, for example copper oxychloride, Mancozeb and Zineb, contain trace elements which contaminate samples collected for leaf analysis and confuse the diagnosis.

**Plant health**  Are plants diseased or infested with insect pests? Pests and diseases can damage the root system or vascular system, which restricts nutrient uptake or mobility to plant tissues causing nutritional symptoms and affecting the leaf analysis. Some produce injuries to tissues that could be confused with symptoms of nutrient disorders.

## Step 2 –
# Diagnosis from visible symptoms

Visible changes in a deficient plant, such as leaf yellowing, stunting of leaves and abnormal fruit development, all begin as a breakdown in the normal metabolism of plant cells. For example, boron deficiency causes death of growing points and distortion of leaves and flowers, because boron is needed for the proper regulation of dividing cells. Similarly, the leaves of nitrogen- or magnesium-deficient plants are pale or yellow because nitrogen and magnesium are constituents of the green pigment, chlorophyll. Such links, between an element's physiological function in plants and a specific abnormality which results when it is deficient, are common in plants. For this reason, the nature of the symptom can provide a useful guide to the identity of a nutritional disorder even in unfamiliar crops.

The two most important diagnostic features of a nutritional symptom are where the symptom is found on the plant (location) and its appearance (colour and pattern).

**Description of symptoms**  Having previously described the general background of the problem (field description), it is now necessary to look closely at the symptoms on individual plants. This is best done in subdued light as occurs in early morning or late afternoon, or ideally on an overcast day. In full sun, light scattered from the surface of the leaf tends to obscure some of the more subtle effects.

**Location**  Where do the symptoms first appear on the plant, for example, on young, mature or old leaves, or in the fruit, seed or bark? Nutritional symptoms generally do not develop uniformly over a plant but show first in specific organs such as the leaves, fruit, roots, shoot or growing point. Leaf symptoms, the most widely used diagnostic feature, can occur in the upper, middle or lower sections of a shoot, depending on the mobility of the element. Mobile elements like nitrogen, magnesium or potassium are moved about the plant relatively easily to satisfy local shortages, particularly in new shoots or developing seeds.

When one of these mobile elements is deficient, the **older leaves** are the first to be depleted and first to show symptoms.

Less mobile elements like iron, boron or calcium, do not move readily from old to younger tissues, so when they are deficient the symptoms appear in the **newer or upper leaves**, the flowers, fruit or seed. Even a temporary shortage of one of these immobile nutrients causes the young tissues growing at the time to suffer and develop symptoms.

Symptoms of nutrient toxicity generally show first in the oldest leaves. These leaves have the highest transpiration rates and consequently receive most of the absorbed nutrients which move in the transpiration stream from the roots.

**Pattern**  Is the leaf uniformly pale in colour or is the chlorosis weakly or strongly patterned? Does the chlorosis or necrosis (burn) begin at the leaf tip, spread around the margin or commence in interveinal areas? Examine the pattern of chlorotic (pale) or necrotic (burnt) areas in relation to vein pattern, and the progress of symptom development. Record the shape and size of necrotic spots or patches on the leaf, the stunting, cupping or downward rolling of leaves, shortened stem internodes, distorted fruit or other tissues, or tissue breakdown. Also note any irregular shape, splitting, cracking or corkiness of affected organs, especially fruit. All of these may help to establish the identity of the disorder.

**Deficiency or toxicity?**  The first question to answer is: Do the symptoms indicate a deficiency or a toxicity? The following generalisations may help to answer this question.
- Deficiency symptoms typically appear on a single leaf-age class (young, middle or old) unless, of course, there is more than one problem.
- Toxicities commonly produce burnt or necrotic (dead) areas of tissue. When new leaves are affected, it almost always indicates a toxicity (except for tip burn caused by calcium deficiency). Burns or necrotic spots on old leaves may or may not be due to toxicity – some deficiencies, notably those resulting from a severe shortage of potassium or magnesium, can also produce necrosis on older leaves.
- Toxicity symptoms often develop rapidly. The affected leaf tissue may change from healthy green to grey-green or dark brown without a transitional yellow phase.
- Symptoms on both old and new leaves usually indicate a toxicity. When an excess of one element causes a nutrient imbalance, deficiency symptoms may be seen in the young leaves while older leaves may show burn or other symptoms of toxicity. Excess phosphorus, manganese or zinc can induce the chlorosis of iron deficiency in young leaves as well as symptoms of nutrient excess in the old leaves.
- Toxicity symptoms usually appear suddenly and may rapidly worsen with time. Strong symptoms are often apparent as early as a day after the plant has been stressed.

**Which deficiency?** If a deficiency is suspected, the location of symptoms on the plant is a useful guide to whether the responsible element is mobile or immobile. The way symptoms develop on leaves, shoots, flowers or fruit provides further clues. Small, irregularly shaped leaves, shortened internodes, aborted flowers, poor seed set and distorted fruit can be characteristic of a particular deficiency.

**Table 2  Quick guide to nutrient deficiencies – what to look for**

**Symptoms first seen in *older* leaves**

*Leaf coloration even over whole leaf*

| | |
|---|---|
| **Nitrogen** | Pale green to yellow leaves. |
| **Phosphorus** | Leaves dull, lacking lustre, bluish green or purple colours. Poor growth. |

*Leaf coloration forms a definite pattern*

| | |
|---|---|
| **Potassium** | Scorching and yellowing, commonly around the edges of leaves, which may become cupped. |
| **Magnesium** | Patchy yellowing often with a triangle of green remaining at the leaf base. Sometimes brilliant red or orange patterns or scorching between veins. |

**Symptoms first seen in *young* leaves**

*Leaf coloration forms a pattern*

| | |
|---|---|
| **Iron** | Almost total loss of green between veins, leaving faint green 'skeleton' of veins on leaf. |
| **Zinc** | Severe restriction of leaf size or stem length, or both (hence the terms 'little leaf' or 'rosetting'). Distinct interveinal creamy yellow patches on leaves in many species. |
| **Copper** | Tip leaves cupped, narrow, distorted or scorched. Defoliation from tip. Chlorosis interveinal or irregular. |

**Symptoms first seen in either *old* or *young* leaves.**

*Leaf coloration forms a pattern*

| | |
|---|---|
| **Manganese** | Mottled diffuse pale green to yellow patches between veins. No restriction of leaf size (unlike zinc). |

**Symptoms usually most prominent in other tissues – seen first in youngest tissues and fruit**

| | |
|---|---|
| **Calcium** | Breakdown of parts of fruit, or death of the growing point. |
| **Boron** | Internal cracking or breakdown of fruit. Irregular shape, corkiness, surface cracking, or gumming in fruit. Irregular flower development or poor pollination and seed set. |

Diagnostic keys, similar to the one in Table 2, provide a framework for a visual diagnosis.

Although visual symptoms are very helpful for diagnosis, the approach does have three major weaknesses:
- Clear visual symptoms do not usually appear until a disorder is quite advanced and some loss of yield or quality has occurred. By this stage, even prompt remedial action will not restore the loss.
- Absence of symptoms in a crop does not mean that nutrition is adequate. 'Hidden hunger' is the condition where performance is limited but no symptoms have been expressed.
- Visual symptoms can be unreliable when more than one element is limiting or when some environmental stress like drought, cold or waterlogging, has altered the normal symptom pattern.

# Step 3 –
# Confirming the diagnosis

Where there is uncertainty about the visual diagnosis, or where an incorrect diagnosis would prove costly, other techniques can be used to confirm the diagnosis.

**Trial treatment of a portion of the crop**  This is the most direct method of confirming a diagnosis. Providing some part of the crop is left untreated to gauge treatment effectiveness, this method gives the most certain answer. But trials can be slow and, if unsuccessful, provide no new information which could lead on to a correct (successful) diagnosis.

**Plant tissue analysis**  Tissue analysis is a powerful diagnostic tool which provides good direct evidence of the nutritional status of a crop. It can be used to verify a visual diagnosis and, because it tests for other nutrients, it enables a new interpretation if the original symptom-based diagnosis is wrong. However, plant analysis may not reveal the underlying cause of a disorder.

> **Plant analysis is used to**
> - Diagnose nutrient deficiencies and toxicities:
>   - confirm a diagnosis based on symptoms,
>   - identify 'hidden hunger', and
>   - suggest additional tests to identify a problem.
> - Predict nutrient disorders in current or future crops.
> - Develop and adjust fertiliser programs.
> - Measure the amounts of nutrients removed in crop produce and residues with the view to replacement.
> - Survey the nutrient status of a crop throughout a district.
> - Compare the nutritional status of soils or growing media.
> - Estimate the dietary value of a crop or pasture.

**Soil and water analysis** These are useful aids in determining the cause of crop nutrient problems but, by themselves, are of limited value in identifying the disorder. Although soil analysis is the only practical means of forecasting crop nutrient needs prior to planting, the test can be misleading if it is not calibrated for both the particular crop and the soil. For example, a soil test calibrated for wheat or for pasture could be misleading if used for a fruit crop. Likewise, quick sap tests of leaves or other tissues which have not been correctly calibrated for a particular crop will not give a reliable guide to that crop's nutrient status.

As no one diagnostic procedure is entirely satisfactory, the best approach is to use a number of techniques to develop a diagnosis which is supported by both laboratory tests and field observations.

## Step 4 –
# Correcting the problem

Corrective treatments should always aim to solve the underlying cause of a problem and not simply reduce the immediate effects on the crop. For example, liming the soil would not be the best way of correcting a calcium deficiency caused by excessive use of potassium fertilisers. In this instance, a series of calcium sprays could be used to minimise damage in the current crop, but long-term correction of the problem will only occur when potassium usage is moderated.

## Step 5 –
# Following up

The only sure way of knowing whether a correct diagnosis has been made, and improving diagnostic skills is to see whether a crop has responded to treatment. Unless the real reason for a crop disorder is discovered, the basic problem may continue to worsen or reappear at some later time and the grower will be no better equipped to deal with it.

## CHAPTER 3

# PLANT ANALYSIS

Plant analysis is the chemical testing of plant tissues (usually leaves) to measure how effectively elements have been taken up by the plant. It can be used to confirm or reject a symptom-based diagnosis, and to determine whether plant growth is limited by nutrition. Plant analysis tells us how well the crop is currently supplied with nutrients but not whether supplies will be adequate in the future. It, therefore, differs from soil analysis which is used primarily to predict the soil's ability to supply the nutrient needs of the crop.

The nutrient status of a crop is assessed by comparing nutrient concentrations in a tissue sample (usually leaves) with predetermined levels or ranges (leaf standards) which have been set for healthy, productive crops of the same species (see Appendix). One criterion often used to decide whether an element is adequately supplied to a crop is the critical level. It is the concentration found when shortage of a nutrient just begins to limit growth (or quality), that is 'the tissue concentration associated with a 10% reduction in yield' (Figure 2). As plant performance can be reduced by toxicity as well as by deficiency, both upper critical (toxicity) levels and lower critical (deficiency) levels can be established for most nutrients.

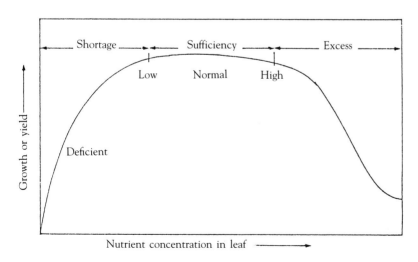

Figure 2 Relationship between plant growth or yield and leaf nutrient concentrations

Leaf standards can also be given in terms of a set of ranges which classify leaf levels as deficient, low, normal, high or toxic. Such ranges better describe the nutritional status from deficiency through normality to toxicity than is achieved with a single critical value. They can also indicate the likelihood of symptoms or quality loss occurring in the crop. High quality of the harvested fruit or its storage quality are often just as important for successful fruit and nut production as top yields and rapid growth.

The five classes of plant nutrient status normally defined are:
- **Deficient**  Symptoms present – nutrient is too low for optimum performance.
- **Low**  No symptoms but nutrient may be too low for optimum crop performance and quality.
- **Normal**  No symptoms – level is adequate.
- **High**  No symptoms – level higher than necessary which may cause imbalance or loss of quality.
- **Excess**  Level too high for optimum performance – toxicity symptoms may be present.

The nutrient composition of a leaf changes during the season and varies with its age and position on the shoot, especially its proximity to developing fruit. The growth stage of the plant (vegetative, flowering, or carrying a heavy crop of fruit) also influences the chemical composition of leaves. All these factors must be accounted for in the leaf analysis procedure. Taking the correct type of leaf at the right stage of the crop's growth is therefore critical in using plant analysis; **leaves must be sampled correctly or the interpretative standards will not apply and the analytical results will be misleading.**

# Sampling method

Procedures for sampling individual crops are given with the nutrient standards in the Appendix. Where these are not available, the following general principles may be used.

**Tissue**  Use whole leaves, that is, blade (lamina) including midrib and petiole (leaf stalk), but extended petioles are often removed. Sample only clean healthy plants of normal growth characteristics.

**Leaf age**  A fully expanded, recently matured leaf.

**Number of plants**  Take leaves from about 25 plants throughout a uniform and representative section of a block within one soil type, following a zig-zag pattern within the sampled area.

**Number of leaves per sample**  60 to 100

- Collect samples from trees that are average for that particular planting.
- Avoid areas shaded by windbreaks or affected by poor drainage or exposed to winds.

- Avoid collecting leaves that are torn, damaged by insects, diseased, stunted, or contaminated by dust or soil.
- Samples should be collected from the same variety and age of tree. If there is a distinct difference of tree health or growth in the same variety, then separate samples from **healthy** and **unhealthy** plants should be collected.
- If there are distinctive soil types in an orchard, collect separate leaf samples from each area.

# Making a two-sample comparison

For many crops there are no standards or the standards are incomplete. Where no leaf standards exist, or their use is inappropriate because sampling could not take place at the correct time or in the specified manner, a comparison of nutrient levels in leaves of healthy and affected plants will help in diagnosing a nutritional disorder (see Table 3). Both the plants and the leaves in both samples must be at the same stage of development. This type of two-sample comparison is also useful for short-lived annual crops to minimise some of the difficulties associated with their rapid maturation and the day-to-day effects of cultural treatments such as irrigations and fertiliser side-dressings.

# Quick sap tests

For vegetable and other annual crops which mature rapidly, requiring nutrient treatments to be determined and applied promptly, quick sap tests are sometimes used for diagnosis. The nutrient composition of the sap can be either laboratory tested or more roughly estimated in the field by the grower using a colour reaction in a tube of solution or on a treated paper strip.

In the latter method, sap squeezed from the petiole of a selected leaf is applied to a strip of paper which has been impregnated with chemicals sensitive to the particular nutrient to be tested. Nitrate is the most commonly used test but tests for some other nutrients including potassium are also available. The concentration of nutrient in the sap is estimated from the intensity of the colour which develops on the test strip and the results are interpreted using standard ranges established for healthy plants. Sap testing is very sensitive to the current nutrient supply but is also very volatile, being subject to many extraneous factors including the time of day when the sample is taken, shading of the sampled leaf, an overcast sky, or the time since the last irrigation. Soluble nutrients quickly rise and then fall after a recent fertiliser side-dressing and can be influenced by drought, irrigation schedule or another nutrient disorder. A quick tissue test usually measures only a single element and can be misleading if another nutrient is deficient or toxic.

**Table 3 Some common deficiencies and toxicities affecting fruit and nut crops**

| Crop | Problem | Occurrence in NSW | Symptom | Nutrient level in leaf | |
|---|---|---|---|---|---|
| | | | | Affected plant | Normal plant |
| Banana | Potassium deficiency | Most districts | Leaf yellowing and scorch on the margins of older leaves with whole leaf finally turning yellow. | 1.6% K | 3.5% K |
| | Magnesium deficiency | North Coast | Yellowing and scorch on leaf margins, stunted growth and reduced production. | 0.08% Mg | 0.3% Mg |
| | Zinc deficiency | North Coast | Stem shortened producing a rosette of chlorotic leaves on very stunted plants. | 12 ppm Zn | 30 ppm Zn |
| Macadamia nut | Potassium deficiency | Most soils especially kraznozems | Light brown necrosis, usually along both leaf margins, or as patches between veins of lower leaves. | 0.12% K | 0.73% K |
| | Magnesium deficiency | Acid soils especially kraznozems and sandy soils | Interveinal yellowing of older leaves, spreading inward from the margins and tip, leaving bands of green tissue on each side of the midrib and between veins. | 0.03% Mg | 0.10% Mg |
| | Nitrogen deficiency | Most soils if fertiliser use low or after a very wet period. | Trees pale and growth poor. Leaves pale and small. | 0.85% N | 1.41% N |
| | Manganese toxicity | Acid soils especially kraznozems, most districts. | Yellowing along the margins of older leaves, especially of the distal half of leaves. Dark brown spots develop within the chlorotic areas. Often combined with magnesium deficiency. | 5460 ppm Mn | 720 ppm Mn |

**Table 3** Some common deficiencies and toxicities affecting fruit and nut crops (continued)

| Crop | Problem | Occurrence in NSW | Symptom | Nutrient level in leaf — Affected plant | Normal plant |
|---|---|---|---|---|---|
| Custard apple | Nitrogen deficiency | Most districts if fertiliser use is low. Soils low in organic matter. After periods of heavy rain. | Foliage pale green and small, with poor shoot growth. Oldest leaves turn yellow and fall. | 1.50% N | 2.73% N |
| | Potassium deficiency | Most soils, especially those with long cropping history. Commonly in autumn after summer flushes and heavy crop load. | Interveinal chlorosis beginning at the margins of older leaves. Black spots and/or brown scorch develop in the chlorotic areas, especially near the margins. | 0.26% K | 1.53% K |
| | Magnesium deficiency | Acid soils, red basaltic (kraznozem) and light sandy loams. | Older leaves develop two bands of interveinal yellowing along each side of the midrib, leaving a green border around the margins. | 0.10% Mg | 0.38% Mg |
| | Phosphorus deficiency | Virgin coastal sandy soils and kraznozems. | Symptoms rarely seen in field-grown crops. Retarded growth, dark blue-green foliage. | 0.07% P | 0.21% P |
| | Zinc deficiency | Most soils including kraznozems. Accentuated by liming. | Young leaves small and narrow with pale green chlorosis between veins. When severe, stunted leaves rosette at the ends of shortened shoots. | 6 ppm Zn | 35 ppm Zn |
| Lychee | Nitrogen deficiency | Most soils where fertiliser use low or after a wet period. | Trees pale, growth poor, leaves uniformly pale green. | 0.90% N | 2.10% N |
| | Potassium deficiency | Most soils, especially heavily cropped light textured or kraznozem soils. | Marginal necrosis of older leaves | 0.44% K | 0.82% K |

**Table 3  Some common deficiencies and toxicities affecting fruit and nut crops (continued)**

| Crop | Problem | Occurrence in NSW | Symptom | Nutrient level in leaf Affected plant | Normal plant |
|---|---|---|---|---|---|
| Guava | Potassium deficiency | Most soils, especially those with a long cropping history. | Light brown marginal burn of older leaves. | 0.51% K | 1.54% K |
|  | Nitrogen deficiency | Most soils where fertiliser use low. | Poor tree growth and pale colour. Leaves uniformly pale green. | 1.3% N | 2.13% N |
|  | Chloride toxicity | Soils of tidal estuaries. Use of saline irrigation water. | Brown scorch along margins of older leaves. | 2.90% Cl | 1.42% Cl |
| Avocado | Nitrogen deficiency | Where fertiliser use is low or after heavy rains. | Leaves very small and yellowish green. If severe, leaves burn and fall prematurely. | 1.02% N | 1.81% N |
|  | Potassium deficiency | Light textured soils and kraznozems. | Irregular-shaped light brown to dark grey necrotic areas between the veins and along the margins of older leaves. | 0.22% K | 1.06% K |
|  | Chloride toxicity | Crops grown in saline soils or irrigated with saline water. | Light grey-brown scorch, from the tip and margins of older leaves, spreading back along the leaf in irregular pattern, unrelated to lateral veins. | 1.37% Cl | 0.05% Cl |
| Kiwifruit (Chinese gooseberry) | Nitrogen deficiency | Most soils if fertiliser use is low. | Stunted growth. Leaves pale green, becoming yellow, especially the older leaves. The veins tend to remain green. | 1.22% N | 2.80% N |
|  | Potassium deficiency | Most soils especially sandy soils and kraznozems. | Marginal and interveinal chlorosis and upward rolling of the edges of older leaves. Light brown necrosis of the chlorotic areas and leaf cupping. | 0.28% K | 2.53% K |
|  | Chloride toxicity | Crops irrigated with saline water or exposed to coastal salt-laden winds. | Older leaves become a dull bluish green colour and curl (upward or down). Dark grey burn develops around the margins. Growth is stunted. | 3.4% Cl | 0.8% Cl |

As yet, sap testing is less often applied to perennial fruit crops than to annuals. This is for a number of reasons including difficulty in squeezing sap from the leaf petioles of many fruit crops and, commonly, the leaves of fruit trees do not accumulate nitrate.

# Adjusting the fertiliser program

Growers can build up a useful record of crop response to fertiliser by monitoring changes in the leaf nutrient levels in a representative block over a number of seasons in relation to previous fertiliser applications, fruit yield and quality, as well as general tree health and vigour. Because crop load, pruning and weather cause large season-to-season effects, trends in leaf levels over several seasons give the clearest picture of the true nutritional status of the crop.

Any adjustments to an existing fertiliser program following a leaf analysis report should be cautious, well considered and, where possible, undertaken in association with sound impartial advice.

It is unwise to make large changes in a fertiliser program on the basis of a single year's leaf analysis results unless the finding is supported by symptoms or other information. A low (or high) value could be a seasonal aberration caused by drought, hard pruning or a heavier than usual crop load. Fluctuations in leaf nitrogen, potassium, calcium and some other elements are common in orchards with alternate bearing trees even when they are well fertilised. High levels of zinc, copper or manganese in leaf samples should also be regarded cautiously, as these elements are present in some nutritional or fungicidal sprays and could be simply a contaminant, even in washed leaves.

# CHAPTER 4
# COMMON NUTRITIONAL PROBLEMS AND THEIR CORRECTION

**N**itrogen deficiency occurs in most soils and most fruit crops need fertiliser nitrogen to achieve acceptable levels of production. The other common nutritional disorders of tropical fruit crops in Australia are deficiencies of potassium, magnesium, calcium, phosphorus, zinc and boron, and toxicities of chloride and manganese.

Soil acidity is endemic to many soils in high-rainfall tropical regions and the problem can be accentuated by cultivation, irrigation and the use of some fertilisers. This can lead to manganese toxicity or, deficiencies of magnesium and calcium. Other nutritional disorders result from a lack of balance between the nutrient elements, or over-fertilising particularly with nitrogen or potassium.

Table 3 lists some common nutritional problems, the soils in which they are likely to occur, the crops most commonly affected, and the levels of nutrients found in the leaves of affected and normal plants.

Records of leaf analyses presented in this book provide an interesting perspective on the relative sensitivity to nutritional disorders of a number of tropical fruit and nut crops. Table 4 shows the percentage occurrence of the main nutritional problems identified from leaf analyses in the period from 1957 to 1988. For example, of the 2500 banana leaf samples received for diagnosis, nutritional disorders were subsequently identified in 25%. The most common problems were potassium deficiency (20% of all disorders), magnesium deficiency (9%), low zinc (23%) and low calcium (26%).

An examination of the frequencies of nutrient disorders over time is also revealing. Distinct differences are apparent in the proportions of certain disorders diagnosed in bananas in the mid 1960s when compared with figures from a decade later (Table 4a). Most notable, are differences in the incidences of two of the most easily recognised and relatively simply corrected problems, namely magnesium and potassium deficiencies. Magnesium deficiency represented 49% of the affected samples in the years from 1965 to 1967 but only 4% in the year 1979/80, while the incidence of potassium deficiency fell from 25% to 8% of the affected samples analysed.

Table 4 Occurrence of nutritional disorders in major tropical fruit crops established from NSW Agriculture plant analysis records 1957–88

| Crop | Percentage occurrence[1] of each nutritional disorder | | | | | | | | | | | Total number of samples | Percentage of samples affected |
|---|---|---|---|---|---|---|---|---|---|---|---|---|---|
| | N | P | K | Ca | Mg | Mn | Cu | Zn | B | Cl | Mn | | |
| | deficiencies | | | | | | | | | | toxicities | | |
| Banana | 11 | <5 | 20 | 26 | 9 | 0 | <5 | 23 | <5 | <5 | <5 | 2542 | 25 |
| Macadamia | <5 | 8 | 9 | 6 | 8 | 0 | 6 | 39 | 5 | <5 | 8 | 201 | 59 |
| Custard apple | 9 | 16 | 27 | 0 | 15 | 0 | <5 | 23 | 0 | <5 | <5 | 128 | 24 |
| Lychee | 8 | 16 | 30 | 0 | 0 | 0 | 0 | 0 | 8 | 0 | 0 | 44 | 27 |
| Kiwifruit | 5 | 6 | 13 | 0 | 0 | <5 | 5 | <5 | <5 | 63 | 0 | 391 | 24 |
| Avocado | 29 | 0 | 16 | 6 | 0 | 0 | <5 | 7 | <5 | 16 | 24 | 754 | 15 |
| Passionfruit | 18 | 14 | 14 | 0 | 7 | 10 | 5 | 5 | 0 | 19 | 8 | 174 | 21 |

[1] Percentage occurrence = $\dfrac{\text{Number of samples with a specific disorder}}{\text{Total number of affected samples}} \times 100$

Table 4a Changes in time of the incidence of nutritional disorders in bananas as reflected in diagnostic leaf analyses for the years 1965–67 and 1979–80

| Period | Percentage occurrence of each nutritional disorder | | | | | | | | | | | Number of samples | Percentage affected |
|---|---|---|---|---|---|---|---|---|---|---|---|---|---|
| | N | P | K | Ca | Mg | Mn | Cu | Zn | B | Cl | Mn | | |
| | deficiencies | | | | | | | | | | toxicities | | |
| 1965–67 | 2 | 2 | 25 | 12 | 49 | 0 | 0 | 5 | 0 | 2 | 3 | 93 | 54 |
| 1979–80 | 9 | 5 | 8 | 46 | 4 | 0 | 4 | 24 | 0 | 0 | 0 | 188 | 36 |

It is believed that these trends reveal an effect of knowledge gained from a decade of diagnostic leaf analyses in the 1960s, enabling farmers and advisory officers to recognise these two disorders and reduce their frequency through modified fertiliser practices. It is worth noting that the frequencies of low calcium and low zinc levels increased with time. This may reflect the fact that slightly low leaf levels of these two nutrients are not always expressed in spectacular leaf symptoms in the plantation, as occurs with deficiencies of magnesium or potassium. Other facts could also be involved, such as soil fertility changes including soil acidification or decline in the organic matter content of the soils.

# Deficiencies

## NITROGEN

Nitrogen is needed in large amounts compared to most other nutrients and nearly all orchard soils require additions as fertilisers or manures. A nitrogen shortage reduces tree growth, leaf cover, blossom formation, fruit set and fruit size.

**Function** Nitrogen is an essential constituent of protein and chlorophyll, the green pigment in leaves. It is quite mobile and the needs of younger leaves, flowers, fruit and new growth can be satisfied from reserves in older tissues. As a consequence, symptoms of deficiency show first, and are most severe, in older leaves.

**Symptoms** All foliage tends to be paler than normal with older leaves becoming yellow and finally dying prematurely. The leaves are small and shoot growth is poor and the thinly foliaged shoots may die back. Tree growth is retarded and crop yield suffers dramatically, through poor fruit set and small fruit size. Fruit colour may be affected. For example, the fruit of nitrogen-deficient lychee has a deep red skin colour, and the midribs, petioles and leaf sheaths of deficient banana plants are pinkish in colour.

Too much nitrogen can promote excessive vegetative growth, often delay fruit ripening, and sometimes result in soft, poorly coloured fruit with poor storage qualities. Banana plants which are over-fertilised with nitrogen appear a rich dark green colour, but the plants are soft, weak and subject to disease, drought and wind. The fruit weight of bananas is actually reduced by excess nitrogen; they are thin, mature later, are soft and have a poor shelf life. The fruit from mango trees given too much nitrogen are prone to 'soft-nose', a breakdown disorder associated with low calcium uptake by the fruit. Reduced nitrogen supply in autumn also increases flowering and fruiting in mangos. An excess of nitrogen in late summer can seriously reduce fruit yield in lychees, by encouraging excessive vegetative flushing in autumn and delaying winter dormancy. Low nitrogen in autumn, winter and early spring is essential for this crop, to ensure good flowering and fruit retention in early spring.

**Corrective measures** If symptoms of nitrogen deficiency are apparent, ensure that waterlogging, drought, or root disease are not responsible for the yellowing before deciding whether the fertiliser program has been adequate. Application rates of nitrogen fertiliser for fruit crops vary greatly (Table 5) from less than 200 g to more than

**Table 5** Suggested mature[1] tree/vine rates (kg/ha/year) of fertiliser for some tropical fruit and nut crops and ranges (in brackets) commonly recommended for a variety of soil and climatic conditions

| Crop (plants per ha) | Nitrogen | Phosphorus | Potassium |
|---|---|---|---|
| Banana (750–1850) | 120 (100–400) | 60 (0–150) | 200 (50–400) |
| Custard apple (120–200) | 100 (50–200) | 30 (10–50) | 100 (50–250) |
| Guava | 100 | 40 | 120 |
| Lychee (70–120) | 100 (60–150) | 40 (30–50) | 120 (80–150) |
| Mango (123) | 100 (40–200) | 25 (0–40) | 100 (60–100) |
| Pawpaw (2000) | 150 (80–350) | 50 (20–150) | 100 (50–250) |
| Pineapple (15 000) | 300 (200–600) | 30 (0–100) | 200 (80–500) |
| Macadamia (200–312) | 100 (60–150) | 25–100[2] | 80 (40–100) |
| Avocado[3] (80–150) | 100 (60–250) | 85 (50–120) | 150 (70–320) |
| Kiwifruit[3] | 170 (150–200) | 60 (20–80) | 100 (80–200) |
| Passionfruit[3] | 400 (200–800) | 50 (20–100) | 150 (50–500) |
| Persimmon[3] | 120 (90–180) | 30 (0–50) | 100 (60–130) |

[1] For young trees or vines use approximately 30–40% or 15–20% of the mature tree rates for 4-year-old or 2-year-old trees respectively, spread around the drip area. Divide the calculated amount into at least four dressings during the main growth months.
[2] Use the lower rate for sandy soils and the higher rate for acid red basaltic clay loams.
[3] Also grown in temperate or subtropical regions, but major plantings are located in the tropical zones of Australia.

1 kg N/tree (that is, less than 50 kg to more than 200 kg N/ha), depending on the crop, soil, rainfall, tree age, spacing, cropping level and the soil management system, for example sod culture. A legume cover crop can supplement the nitrogen supplied from fertilisers but a grass sod may demand additional nitrogen fertiliser especially in the early years of its establishment.

Common sources of fertiliser nitrogen include ammonium nitrate, urea, ammonium sulphate, N–P–K fertiliser mixtures, or manures such as poultry manure. Table 6 lists some common fertilisers and manures, showing the quantities to be applied to supply 1 kg N, P or K. As most soils on which tropical fruit are grown are acid and nitrogen fertilisers

Table 6  Some commonly used fertilisers for tropical fruit and nut crops

| Fertiliser | Nutrient content | kg of fertiliser needed to supply 1 kg of N, P or K | Advantages | Problems |
|---|---|---|---|---|
| Urea | 46% N | 2.17 | High analysis – low freight, low cost. | Losses if left on surface. Water or cultivate in. |
| Ammonium nitrate | 34% N | 2.94 | Low cost. Quick response. Half N as nitrate which moves quickly to the roots. | Nitrate can leach easily. |
| Ammonium sulphate | 21% N | 4.76 | Not recommended for orchards. | Expensive source of nitrogen. Highly acidifying. |
| Diammonium phosphate (DAP) | 18% N 20% P | 5.55 for N 5.00 for P | Supplies both N and P. | Strongly acidifying. Concentrated source of N and P but lacks calcium. |
| Single superphosphate | 8.8% P | 11.36 | Supplies P, Ca and S for both sod and trees. | Relatively low analysis. |
| Double superphosphate | 17% P | 5.88 | Concentrated source of P giving savings in freight and handling. | |
| Potassium chloride (muriate of potash) | 50% K | 2.00 | Cheapest form of K. Highly soluble. | Has a high chloride content. Do not use where soil salinity is a problem. |
| Potassium sulphate | 41% K | 2.44 | Free of chloride. | Its higher cost compared to muriate of potash is not usually warranted. |
| Potassium nitrate | 38% K 13% N | 2.63 for K 7.69 for N | Supplies both N and K in soluble form. | Expensive. |
| N–P–K mixtures | Various proportions of N, P and K | – | A convenient way of supplying N, P and K in one application. | More expensive than single element fertilisers. Wasteful unless P and K are needed. |
| Blood and bone[1] | 5% N 4% P | 20 for N 25 for P | Supplies both N and P in long-lasting forms. | Large quantities needed. Expensive. Supplies limited. |

Table 6  Some commonly used fertilisers for tropical fruit and nut crops (continued)

| Fertiliser | Nutrient content | kg of fertiliser needed to supply 1 kg of N, P or K | Advantages | Problems |
|---|---|---|---|---|
| Poultry manure (high grade) | 3.3% N<br>2.0% P<br>1.5% K | 30 for N<br>50 for P<br>67 for K | Nutrients in longer lasting forms. | Large quantities needed. |
| Poultry manure (fair grade) | 2.0% N<br>1.2% P<br>1.0% K | 50 for N<br>83 for P<br>100 for K | As above. | As above. Nutrient content variable. |

[1] Typical analysis: the analysis of blood and bone varies from brand to brand.

differ in the extent to which they increase this acidity, the potential acidifying properties of the fertiliser needs to be considered when choosing the best form of nitrogen for your particular soil and crop. Table 7 shows ammonium sulphate to be very acidifying while potassium nitrate and calcium nitrate (which are usually more expensive) do not acidify at all but have a slight alkaline effect on soil pH.

Foliar feeding can be useful for quickly rectifying a deficiency, or for supplementing normal soil-applied fertiliser especially after losses from heavy leaching rain storms or when an unexpectedly large crop has set. However, uptake from sprays is limited and, except for pineapples, they are not a substitute for soil dressings. Urea is the usual nitrogen fertiliser used in foliar feeding. It is also added to some trace element sprays to enhance uptake. Only low biuret urea (<0.4%) should be used, to avoid phytotoxicity. Urea concentrations as high as 10 kg/100 L are used on pineapples which are particularly resistant to burning, but 0.5 kg/100 L is a safer concentration for most other crops.

Table 7  Potential acidifying properties of nitrogen fertilisers

| Fertiliser | Approximate lime requirement[1] (kg calcium carbonate/kg N used) |
|---|---|
| Ammonium sulphate | 5.5 |
| Monoammonium phosphate | 5.5 |
| Diammonium phosphate | 3.5 |
| Ammonium nitrate | 2.0 |
| Urea | 2.0 |
| Potassium nitrate | −2.0 |
| Calcium nitrate | −2.0 |

[1] For average leaching conditions

## 26 Common nutritional problems and their correction

1

3

2

4

5

Deficiencies 27

6

9

7

8

**1 Custard apple** – the deficient leaf (left) develops an even paleness (that is, without a pattern) over the whole leaf which is also smaller than normal. Leaf N = 1.9% vs 2.8%.

**2 Custard apple** – when nitrogen supply is limiting, growth is stunted and older leaves are the first to become pale and finally turn bright yellow, before falling off prematurely.

**3 Avocado** – small narrow pale leaves which become stiff and slightly inwardly rolled. Tree growth is stunted and fruit production reduced if temporary shortage is not promptly corrected. Leaf N = 0.9%.

**4 Avocado** – increasing paleness and stunting of leaves due to nitrogen shortage (left to right: severe deficiency, mild deficiency and normal healthy leaf). Leaf N (affected) = 1.0%.

**5 Avocado** – when the deficiency is acute (far right) leaves can become an orange-yellow colour, they begin to burn from the tip and finally fall. Leaf N = 0.8%.

**6 Macadamia** – pale green leaves with veins distinctly lighter in colour than the interveinal areas, making them appear more prominent than normal. Leaf N = 1.0%. Slight clearing of the veins may be seen in nitrogen-deficient leaves of a number of fruit crops, but kiwi fruit is an exception (9).

**7 Pineapple** – older leaves are pale and become yellow especially towards the tip and margins. (R. Broadley, QDPI)

**8 Passionfruit** – juvenile leaves of a young vine showing increasing degrees of deficiency, from pale green (centre), to yellow and stunted (right) compared to the healthy leaf (left). Leaf N (left to right) = 4.9%, 2.5% and 1.9%.

**9 Kiwifruit (chinese gooseberry)** – leaves become pale green to yellow with the oldest leaves being most affected. Contrary to the uniform loss of green colour from leaves which characterises nitrogen deficiency in most other plants, the veins of deficient kiwifruit leaves remain green.

# PHOSPHORUS

Many orchard soils have low levels of available phosphorus in their original unfertilised state. Cover crops and clover-based sod often benefit from phosphate application, but reports of mature fruit trees or nut crops responding in yield or growth or displaying deficiency symptoms are less common.

**Function**  Phosphorus is important for cell division and growth. It is needed for photosynthesis, sugar and starch formation, in energy transfer, and for the movement of carbohydrates within the plant.

**Symptoms**  Symptoms and marked growth effects are not often seen in commercially grown fruit trees and vines. Where phosphorus is deficient (young trees on very infertile soils), shoot growth is restricted and leaves are small – often dull bluish green or bronze in colour with tints of purple especially in the veins on the undersurface of the leaf. Macadamia trees on acid red basaltic soils exhibiting sparse foliage and bronze-coloured leaves have responded to liberal applications of superphosphate (500 kg/ha).

Pineapples respond to phosphorus in poor soils. Deficient plants are stunted and root growth is very poor. The leaves are darker than normal and the oldest leaves develop a purple-red colour with yellow margins. Later these leaves senesce and die back. Fruit and sucker production is limited.

Some Australian native plants are particularly sensitive to high levels of soluble phosphorus in the soil or growing medium. Even though macadamias are native to Australia, they respond favourably to generous applications of phosphorus in a well balanced fertiliser program (Table 5).

**Corrective measures**  Superphosphate, ammonium phosphates (MAP or DAP) and compound N–P–K fertilisers are the usual means of supplying phosphorus to orchards. Poultry and animal manures will significantly contribute to soil phosphorus supplies if applied at high rates (>50 tonnes/ha) over a number of years. Because phosphorus from surface-applied fertilisers does not move easily down through undisturbed soil, correction of a deficiency can be slow in established trees. Apply 44–88 kg P/ha ($\frac{1}{2}$–1 tonne/ha single superphosphate) as a strip in the drip area and under the trees.

After many years of regularly using phosphorus fertilisers, reserves can build up in some soils enabling less frequent applications. Too much phosphorus can accentuate zinc and iron deficiencies, especially in low fixing sandy soils.

# Deficiencies

**1 Custard apple** – symptoms of phosphorus deficiency are seldom noticed in the field because they are usually not distinctive in fruit crops and most orchard soils have received generous fertilising in the past. Growth of these pot-grown custard apples is stunted and the leaves are dark green but dull in colour.

**2 Avocado** – plant growth is stunted, leaves are small and new growth is distinctly bronze in colour. Leaf P = 0.09%.

**3 Passionfruit** – old leaves acquire a dark bluish green colour, often with chlorosis adjacent to the main veins at the base of the leaf. Leaf P = 0.05%.

**4 Passionfruit** – young leaves are very stunted and narrow, and tend to roll inward.

**5 Pineapple** – stunted plants with poorly developed roots in a severely deficient patch of a young crop. Older leaves of the affected plants become purplish red with yellow margins and die back from the tips. The young heart leaves often remain green. (R. Broadley, QDPI)

# POTASSIUM

Potassium deficiency is relatively common in tropical fruit and nut crops, particularly bananas, custard apples and macadamias. Bananas are a very high potassium-demanding crop; a mature plant has more potassium in its tissues than any other element including nitrogen. Potassium demand during periods of rapid vegetative growth or bunch filling often exceeds the rate of release from the soil. Many other tropical fruits also have very high seasonal demands for potassium, and deficiencies would be more common than they are but for the regular large applications of potassium fertilisers routinely given to most crops.

The potassium needs of fruit crops are met chiefly from exchangeable soil potassium, which is slowly replenished by mineral reserves, or by fertilisers. In general, sandy soils are less well supplied with potassium than heavier textured soils. Losses through leaching run-off and erosion can be quite significant in tropical climates and hilly terrain. Further losses occur in the crop removed from the farm each year. Drought can reduce potassium uptake by drying surface soil where much of the available potassium usually exists, and waterlogging can lower uptake by restricting root activity.

**Function** Potassium is important for the formation of proteins, carbohydrates and fats, and for the functioning of chlorophyll and several enzymes. It is needed in cell division, for maintaining the balance of salts and water in plant cells and for opening and closing the stomates – the tiny breathing pores on the undersurface of leaves. Potassium is highly mobile and can be moved freely within the plant to new tissues when needed. As a consequence, the older leaves are the first to show deficiency symptoms. Concentrations of potassium are greatest in leaves, growing points, flowers and fruit.

**Symptoms** Yield and quality can be affected by potassium deficiency even before leaf symptoms are seen. Fruit from potassium-deficient plants are often small, poorly coloured and taste insipid. Bunch initiation is delayed in bananas, fruit size and numbers are drastically reduced and the fruit fails to fill.

Scorching of the leaf edge and sometimes between the veins characterises potassium deficiency in many crops. Before leaf burn develops, older leaves may become bluish green in colour. This may be followed by some chlorosis around the leaf edge or between the veins.

In bananas, older leaves turn a golden yellow, beginning near the tip and distal margin. The end of the leaf folds down as the midrib collapses, usually about one-third back from the tip. This decline of the leaf, from the first sign of chlorosis to complete collapse ('leaf fall') is very rapid, often happening within a week. Avocados develop an irregular interveinal chlorosis extending inward from the margins of older leaves. Affected areas are initially yellow-green, but turn tan and then brown before finally scorching. Symptom leaves often curve downwards.

Young potassium-deficient pineapple plants produce narrow dark green, stiff leaves. The leaves later turn yellow and die from the tips. Fruit are small with pale insipid-tasting flesh, low in both sugar and acid.

**Corrective measures** Significant amounts of potassium are removed in fruit and nut crops (20 kg K/ha can be taken off in one crop of bananas or macadamias). Soils have natural reserves in the clay minerals but most soils still need regular potassium fertiliser supplements.

Potassium chloride (50% K) (muriate of potash) is the cheapest and most common fertiliser source of potassium. Potassium sulphate (42% K) contains less potassium and is more costly, and its use is only warranted where salinity is a concern. Potassium nitrate (38% K, 13% N) contains nitrogen as well as potassium. It is expensive but a useful fertiliser for fertigation applications. Mixed N–P–K fertilisers can supply regular maintenance levels of potassium in appropriately balanced combinations.

When potassium deficiency occurs in tree crops, a single heavy dressing of 200–300 kg K/ha applied as muriate of potash may be needed to rectify the deficiency relatively quickly, rather than using normal maintenance rates or N–P–K mixtures. Applications are generally made before peak periods of demand such as growth flushes or fruit filling. Surface banding near the drip ring of the trees improves uptake.

Potassium fertiliser is expensive, and can be harmful if used to excess. Too much potassium can depress calcium and magnesium uptake. To avoid such adverse interactions, magnesium (as magnesite or dolomite) is often given in conjunction with potassium to crops such as bananas and pineapples which require heavy rates.

1

**1 Macadamia** – light brown necrosis develops between the veins or along the margins of mature leaves. Leaf K = 0.10%.

2

4

3

5

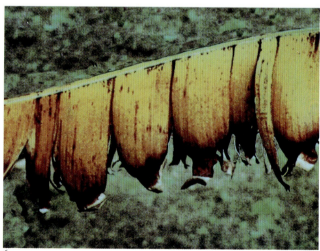
6

Deficiencies 33

7a

**2 Macadamia** – necrotic lower (older) leaves with young leaves towards ends of the shoots showing no symptoms. Potassium is readily mobile and is withdrawn from old leaves, to supply the needs of new growth and nuts, when soil potassium is inadequate.

**3 Custard apple** – In this glasshouse-grown plant, a mild interveinal chlorosis of young (upper) leaves becomes a dark brown to ash grey necrosis in middle-aged leaves. The lowest, first formed, leaves have developed a lighter brown necrosis, less common in field-grown trees.

**4 Custard apple** – yellowing, followed by necrosis from the leaf margin, progresses towards the centre of these mature leaves from a field-grown tree. Leaf K = 0.30%.

**5 Custard apple** – a more severe stage of deficiency where most of the leaf tissue towards the margin and leaf tip has become necrotic, with tissue near the mid vein and leaf base still green but becoming chlorotic. Leaf K = 0.21%.

**6 Banana** – often called 'leaf fall', 'premature yellowing' or 'yellows'. Potassium deficiency often develops suddenly when soil supplies cannot match seasonal demand as fruit bunches develop. Yellowing begins at the tip and margin of older leaves, and then spreads rapidly until the whole leaf becomes a golden yellow colour within 2–3 days.

**7a and 7b Banana** – withering and death of the lamina ('leaf fall') quickly follows chlorosis ('yellows'). The end of the leaf often bends downward and the midrib may break at about one-third of its length from the tip. Leaf K = 0.64%.

**8 Banana** – fruit breaks down following leaf death. When the deficiency is milder, bunches form but fail to fill.

7b

8

9

10a

10b

Deficiencies    35

11

12

13

**9 Avocado** – interveinal chlorosis is often the first sign of deficiency. Light brown spots may appear scattered over the leaf. Leaves may be narrow (left) and slightly smaller than normal. Leaf K = 0.14% (affected) and 1.3% (healthy, at right).

**10a and 10b Avocado** – large irregular necrotic spots develop in the interveinal chlorotic areas (10a) or around the margin (10b).

**11 Avocado** – large necrotic areas can also develop with little prior leaf chlorosis.

**12 Passionfruit** – young leaves (left) can develop a speckled light straw-coloured necrosis, while older leaves (right) burn from the tips after becoming chlorotic between the veins. These older leaves then fall.

**13 Pineapple** – leaves on the young deficient plant are dark green, stiff, narrow and deeply troughed. As the plant ages, leaves of a deficient plant become yellow and die back from the tip. (R. Broadley, QDPI)

36   Common nutritional problems and their correction

14

15

17

16

Deficiencies    37

18

**14 Ginger** – chlorosis and necrosis progressing inward from the margin and leaf tip of older leaves. The leaf margin tends to roll inward from its burnt edge. (G. Sanewski, QDPI)

**15 Kiwifruit** – a pale green interveinal chlorosis develops from the margin.

**16 Kiwifruit** – a later stage of deficiency in this field specimen showing inward rolling of the leaf margin and both marginal and interveinal necrosis preceded by interveinal chlorosis. Leaf K = 0.60%.

**17 Persimmon** – chlorosis commencing around the margins of mature leaves followed by marginal necrosis.

**18 Avocado** – potassium deficiency induced by drought. The leaves have developed the interveinal necrosis of potassium deficiency but are also inwardly rolled, especially towards the tip, due to moisture stress on the tree. Drought can precipitate potassium deficiency by limiting root activity in the surface soil where much of the available potassium from fertiliser and the richer top soil is located.

# CALCIUM

*L*ow soil calcium rarely limits tree growth or the yield of fruit or nuts. Calcium tends to be lower in leached acid soils but the deficiency is not confined to these soils. Many calcium-related disorders are the result of calcium movement patterns within the plant when the supply to young tissues – such as growing points, young leaves, root tips or developing fruit – is interrupted at a critical stage. Conditions which stress the plant – including drought, very hot weather or drying winds, or factors which encourage rapid tree growth such as heavy pruning or excess nitrogen – can all reduce calcium movement into the fruit and other developing tissues. Excessive use of potassium fertiliser will lower calcium uptake from the soil.

**Function** Calcium is an important constituent of the cell walls and membranes. When it is in short supply, cell membranes become leaky, and cell division is disrupted causing abnormalities in the growing points and root tips. Calcium appears important in protecting the cell from toxins, in slowing the aging of plant tissues, and in promoting longer storage life and resistance to tissue breakdown in some fruits.

**Symptoms** Because calcium has poor mobility within the plant, symptoms of deficiency appear first in young rapidly growing tissues such as buds, young leaves, roots and fruit. Breakdown of the flesh of mango fruits has been shown to be related to low calcium and high nitrogen. This condition, described as 'soft-nose', develops on the ventral side and towards the distal end of the fruit while it is still on the tree. The emergence of distorted and stunted young banana leaves – a condition described as 'spike-leaf', 'dog-ear' or 'November leaf' – results from a temporary shortage of calcium; the lamina is deformed or almost absent in affected leaves. Calcium sprays alleviate this condition which usually occurs during rapid growth flushes in spring, in heavily fertilised, productive plantations.

New heart leaves of deficient pineapples also emerge distorted, having scalloped margins and 'cut-off' tips. Another effect is a high incidence of abnormalities in the fruit including joined multiple pineapples ('siamese twins' or 'teapots') or round fruit ('cannon balls').

**Corrective measures** Calcium is supplied in lime and dolomite, often needed to prevent the soils on which most tropical fruit and nut crops are grown from becoming too acid. Most crops prefer a soil which is slightly acid, that is, with a pH of about 5.0–5.5 (1:5 $CaCl_2$) or 5.5–6.0 if the pH is measured in water. A common practice in high-rainfall tropical areas, where the soils are usually naturally acid and fertiliser usage relatively high, is to apply 2–3 tonnes/ha of lime or dolomite every second year. Gypsum applied at 1–2 tonnes/ha, is an alternative source of calcium where soil acidity is not a problem, as it does not appreciably alter soil pH. It is soluble and quickly moves to the root zone in high-rainfall areas.

These treatments should avoid chronic calcium shortage but a temporary shortage as described above (for example 'spike-leaf' in bananas), when calcium supply cannot meet the demands of rapidly growing new tissues, can only be overcome by foliar sprays. Apply calcium nitrate sprays at the rate of 1 kg/100 L (banana leaves) or 500 g/100 L to fruit of other affected crops. Avoid excessive applications of nitrogen or potassium fertilisers.

1a

2

1b

3

**1a and 1b Banana** – symptoms caused by a temporary shortage of calcium during a period of rapid growth. Sections of the lamina fail to develop along parts of the midrib of young emerging leaves. The condition, described as 'spike-leaf' or 'dog-ear' can be accentuated by heavy applications of nitrogen and potassium fertilisers. It is also common in the strong flush of growth which follows a cyclone. (D. Stevenson)

**2 Pineapple** – new heart leaves emerge with irregularly shaped margins, outlined by a narrow chlorosis. The tips of many leaves look as if they have been cut off and the margins resemble the damage caused by chewing insects such as locusts.

**3 Pineapple** – death of the growing point can result from calcium shortage prompting the emergence of secondary buds or high suckers. (D. Swete Kelly)

40  Common nutritional problems and their correction

4a

4b

5

**4a and 4b Pineapple** – a number of abnormalities can occur in the fruit of deficient plants including multiple fruits ('siamese twins' (4a), and 'teapots' (4b)) or rounded fruit. (R. Broadley, QDPI)

**5 Custard apple** – the blossom ends of fruit turn black, both the skin and the flesh beneath it. (G. Sanewski, QDPI)

# MAGNESIUM

Magnesium deficiency most often affects bananas, macadamias and custard apples growing on leached acid soils low in organic matter. Soils formed from sandstone usually have low reserves of this element and quickly become deficient, but the basalt-derived acid red loam (kraznozem) soils also become depleted in highly productive plantations. Bananas remove significant amounts of magnesium in the fruit (about 0.4 kg Mg/tonne), while the relatively large amounts of nitrogen and potassium fertilisers normally used to maintain high yields of bananas and other tropical fruit, further reduce uptake. The balance of magnesium to other soil minerals, particularly potassium and calcium, is disturbed by such continued heavy use of fertilisers and by liming with agricultural lime (calcium carbonate), unless some magnesium (for example dolomite) is included in the fertiliser/liming program. Magnesium deficiencies are more common during dry seasons and when cropping is heavy.

**Function** Magnesium is present in chlorophyll, the vital green pigment which enables plants to form sugars and starches from atmospheric carbon dioxide. It is also important for the functioning of several plant enzymes, particularly those associated with photosynthesis. Magnesium is mobile within the plant and moves from older to newer tissues in times of shortage and, as a consequence, the oldest leaves are the first to show deficiency symptoms.

**Symptoms** Most magnesium-deficient crops develop a characteristic bright yellow interveinal chlorosis in older leaves. This yellowing usually begins at the tip and margins of the leaf and spreads towards the midrib, often leaving a green triangular area near the base of the leaf. However, in custard apples, bananas, and some other crops, chlorosis may not begin at the margin but slightly in from the leaf edge. Yellow bands extend along each side of the lamina, leaving narrow strips of green along both margins and on each side of the midrib. In bananas, these spreading chlorotic bands tend to develop independently of the lateral vein arrangement. Brown necrotic patches can develop within the chlorotic areas of old, severely affected leaves in some species including custard apple, avocado and persimmon. Early defoliation of leaves is common in magnesium-deficient crops.

**Corrective measures** The main materials used to correct a deficiency of magnesium are the liming materials, finely ground dolomite (calcium magnesium carbonate 8–13% Mg), magnesite (magnesium carbonate 20–28% Mg) and magnesium oxide (40–55% Mg) or the soluble salt magnesium sulphate (Epsom salts 9.6% Mg). In established plantations on acid soils, 2–4 tonnes of dolomite or ½–1 tonne of magnesium oxide per hectare every second year will correct magnesium deficiency as well as help reduce soil acidity. However, correction may take two or more years. For replant bananas, placing 250 g dolomite per plant mixed through the soil in the planting hole will give quick correction. Foliar sprays of magnesium sulphate (Epsom salts) can provide a quick but short-term response while soil applications are bringing about a longer term correction. Use two to three sprays of Epsom salts plus urea (1 kg of each/100 L) for custard apples or Epsom salts alone (2 kg/100 L) for persimmons. Better management, such as reducing potassium fertiliser inputs where these are found to be high or using mulches or irrigation to maintain adequate soil moisture during dry seasons, can also improve magnesium supply.

1a

2

1b

**1a and 1b Macadamia** – yellowing from the margins, especially at the tip end of the leaf, tending to retain a 'Christmas tree'-shaped band of green tissue along the midrib and major veins. Light orange-brown necrosis then develops in the chlorotic areas (leaf at right (1a) and in the three on the left of 1b).

**2 Macadamia** – chlorosis develops in older leaves while the youngest leaves of subsequent growth are green.

Deficiencies    43

**3 Custard apple** – interveinal chlorosis develops in bands on each side of the midrib of mature leaves with a green border around the margins (left). Dark reddish brown necrosis then develops within the chlorotic areas especially between the veins (centre and right). Leaf Mg = 0.09%.

**4 Custard apple** – a mature shoot showing some of the range of chlorotic and necrotic patterns which may be seen in deficient leaves.

**5 Banana** – yellow bands of chlorosis spread along the full length of mature leaves, and on both sides of the blade. Note that while the older leaves are affected the younger upper leaves are still green.

**6 Banana** – chlorosis can begin at the margin or a few centimetres in from it, when the leaf will retain a green border for some time. The green strips on each side of the midrib are the last parts of the leaf to lose their greenness. (F. Chalker)

## 44 Common nutritional problems and their correction

7

8

9

**7 White sapote** (casimiroa) – interveinal chlorosis of older leaves with a tendency for tissue along the midrib and near the leaf base to remain green the longest (centre leaf). Leaf Mg = 0.23%.

**8 Passionfruit** – yellow chlorosis of mature leaves with areas near the three main veins being the last to lose their colour. Leaf Mg = 0.10%.

**9 Passionfruit** – the developing chlorosis can often leave a green margin around the leaf as also occurs in custard apple and bananas. Leaf Mg = 0.11%.

**10 Avocado** – interveinal chlorosis progresses inward from the leaf margins and tip towards the leaf base, midrib and lateral veins. Leaf Mg = 0.14%.

**11 Persimmon** – yellowing between the veins of mature leaves. (J. Campbell)

# Sulphur

Sulphur is required by plants in roughly the same amounts as phosphorus. A crop of bananas, for example, can remove from the soil over 15 kg S/ha in the fruit alone. Despite this heavy demand, deficiencies are rare in tree crops or pineapples, presumably because adequate amounts of sulphur are contributed in fertilisers such as superphosphate, ammonium sulphate, potassium sulphate and gypsum. Additional sulphur comes in rainfall, particularly in areas near the sea or industry, and in irrigation water and many horticultural sprays such as fungicides. Cultivation also releases available sulphur from soil organic matter, making deficiency less common in crops than in pastures.

Sulphur deficiency has been reported, however, in field grown bananas in the Windward Islands.

**Function** Sulphur is a constituent of protein, amino acids and co-enzymes and is involved in many critical plant processes. Sulphur has low mobility within the plant so when supply is inadequate, symptoms show first in the youngest tissues.

**Symptoms** With a severe deficiency, the youngest leaves acquire a uniform chlorosis which is accentuated by mature, neighbouring leaves remaining dark green. In some crops, the symptoms can be difficult to distinguish from iron deficiency. However, unlike iron chlorosis, the veins of sulphur-deficient leaves are not green and the leaf blade is a dull milky yellow.

In bananas, the normal greening of emerging leaves is at first delayed, and then stops entirely resulting in creamy white coloured leaves. The margins are the most chlorotic and develop necrotic areas as the leaves age. In extreme cases, new leaves emerge with distorted and reduced leaf blade tissue or remain as little more than bare midribs.

**Corrective measures** Most fertiliser programs supply sufficient sulphur in common fertiliser components. If sulphur deficiency is encountered, 30–50 kg S/ha will meet the needs of most crops. This is equivalent to 300 kg gypsum, 400 kg single superphosphate, 200 kg of ammonium sulphate, 300 kg of potassium sulphate, or appropriate combinations of some of these fertilisers.

Deficiencies 47

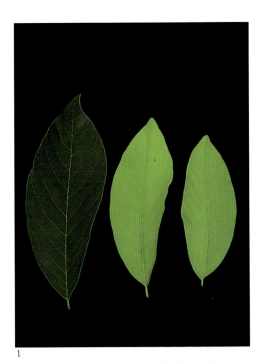

**1 Custard apple** – deficient leaves (centre and right) are uniformly pale, resembling the leaf symptoms of nitrogen deficiency (nitrogen 2), except it is the young growth which is most affected when sulphur is deficient.

**2 Avocado** – pale green upper (young) leaves contrast with normal dark green lower mature leaves. Because sulphur is not very mobile, it does not move freely from old leaves to supply the needs of young growth as happens when mobile nutrients like nitrogen are deficient.

#

*I*ron deficiency is rarely caused through insufficient iron in the soil, but usually because the iron has been rendered unavailable for plant uptake by alkaline soil conditions or an excess of manganese or phosphorus. In many agricultural crops, iron deficiency is a problem of high pH calcareous soils and it is often described as 'lime-induced chlorosis'. These conditions are reported in some banana growing soils of Israel, Hawaii and certain coral-derived soils of ocean islands but are rare in tropical fruit growing areas in Australia where the soils are usually acid to very acid. Despite this, iron deficiency is still a significant problem of pineapples and custard apples in Australia.

Deficiencies in pineapples on acid red basaltic soils, on which they are often grown in southern Queensland and northern New South Wales, are caused by very high manganese. Custard apples appear relatively sensitive, suffering from iron deficiency where other crops such as bananas or avocado are usually not affected. This sensitivity appears to be related to the crop's poor ability to absorb or utilise iron. High soil phosphate is thought to be a factor in some areas. Poor root health may also be important, especially the health of the root tips where much of the iron needs of a plant are absorbed. Accordingly, mulching to protect and encourage healthy feeder roots is a principal strategy in controlling this deficiency in custard apples.

**Function** Plants need iron to produce chlorophyll and to activate several enzymes including those involved in the oxidation/reduction processes of photosynthesis and respiration.

Iron concentrations of 50–100 ppm are often quoted as satisfactory leaf analysis values for most crops, but leaf analysis is not a reliable guide to iron deficiency. There is poor mobility between tissues. High or normal values are frequently found in analyses of clearly iron-deficient leaves, because iron can exist in the leaf in forms which are inactive for plant metabolism yet register in the analysis. Contamination of the leaf surface by dust is a further problem, though it can be overcome by stringent washing procedures, provided the leaves have not wilted prior to washing. Therefore, while a low iron value in a leaf analysis indicates deficiency, a 'normal' or high value does not prove a sufficiency of this element.

**Symptoms** The youngest leaves develop a light green chlorosis of all the tissue between the veins. A distinctive pattern results from the network formed by the midrib and veins which remain green, for example custard apples. If the chlorosis is severe and persistent, yellowing increases to the point of bleaching, and burns can develop within these chlorotic areas. Because iron does not move easily within the plant, older leaves can remain green while flushes of new growth are chlorotic. In pineapples, chlorosis is strongest towards the margins of young inner leaves. The fruit are small, reddish in colour, hard and prone to cracking. Coconuts develop a uniform yellowing of tissue between the parallel veins of leaflets. Mangos are unlike most other crops in that the newest leaves of iron-deficient trees initially turn a uniform yellowish green and then stop growing. The shoot may eventually die back from the tip.

## Corrective measures   (Table 8)

### Foliar sprays
*Pineapples* – 3% ferrous sulphate (3 kg/100 L) applied at low volume (100–500 L/ha). Do not spray past run-off. Urea at the rate of 10 kg/100 L is often added to the spray.
*Bananas and avocados* – Use 0.5% (500 g/100 L) of ferrous sulphate.

### Soil application
*Kiwifruit* – Apply ferrous sulphate 10 g/m² of soil under the canopy (50 g/vine). Iron chelates at 10–20 g/m², can be more effective on heavy neutral to alkaline soils but they are very expensive for treating more than just a few plants.

For container-grown plants in the nursery, drench the medium and leaves with a solution of iron chelate (12% Fe) at the rates of 0.5 g/L or ferrous sulphate at 5.0 g/L.

Table 8   Some commonly used trace element fertilisers for tropical fruit and nut crops

| Nutrient and occurrence | Material and application method | Advantages | Comments |
|---|---|---|---|
| **Iron** Deficiency is usually the result of excessive liming. Also induced by high phosphate usage or high soil manganese. Observed in heavy black soils when drainage is poor. | Iron sulphate (23% Fe) S* | Low cost. | |
| | Iron chelates (5–15% Fe) F* and S* | Versatile for applying foliar or soil applications. | Expensive. |
| **Manganese** Deficiencies can occur after liming, especially in coastal sandy soils. | Manganese sulphate monohydrate (28–36% Mn) F* and S* | Foliar spray gives quick response. | Not usually a problem in coastal acid soils. |

**Table 8  Some commonly used trace element fertilisers for tropical fruit and nut crops (continued)**

| Nutrient and occurrence | Material and application method | Advantages | Comments |
|---|---|---|---|
| **Zinc** Low to deficient levels occur in many coastal soils. | Zinc sulphate heptahydrate (23% Zn) F* and S* | Soluble, used for foliar sprays in most tree crops. | Can be neutralised with fresh hydrated lime to reduce the risk of burn in some crops. |
| | Zinc oxide (60–80% Zn) S* | Cheapest and most concentrated form of zinc. Absorbed more slowly by leaves avoiding leaf burn and other injuries. | Difficult to spread in small quantities. Particles failing to pass through a 60 mesh sieve are of little use as soil or spray treatments. |
| | Zinc sulphate monohydrate (36% Zn) S* | In prilled or granular forms easy to apply through fertiliser spreaders. | Does not dissolve easily and not favoured as foliar spray. Effective when cultivated into the soil. |
| | Zinc chelates (5–15% Zn) F* and S* | Soluble in water. More mobile in soil than other forms. | More expensive and less concentrated than other forms. |
| **Copper** Low to deficient levels occur in coastal sandy or peat soil. | Copper sulphate (bluestone) (25% Cu) S* | Cheap form of copper for soil dressing. | Fine powdered forms required to dissolve quickly. Mix with fresh hydrated lime in equal parts for foliar Bordeaux mixture spray. |
| | Copper oxychloride (55% Cu) F* | Cheap source of copper for spraying. | Usually a safe spray but can injure sensitive foliage and fruit of some crops. |
| **Boron** Acid leached granitic or sandy soils low in organic matter or heavily limed soils. Dry conditions following a wet spell often leads to deficiency. | Borax (11.3% B) S* | Used as a soil dressing. | Difficult to dissolve for foliar sprays. |
| | Polyborate powder (20.5% B) F* and S* | Soluble and compatible with many insecticides, fungicides and herbicides. Double strength of borax. | Incompatible with oil-based solutions. |

F* = Foliar spray.
S* = Soil application.

Deficiencies   51

**1 Custard apple** – bright yellow chlorotic young leaves on a new shoot. Iron chlorosis is distinctive, with only the veins remaining green and usually being confined to patches of new growth on several parts of the tree. This differs from nitrogen deficiency, for example, where the chlorosis is spread throughout the tree, the oldest leaves are palest, and the veins do not remain green.

**2 Custard apple** – chlorosis affects all of the interveinal leaf tissue of the youngest leaves with only a net-like pattern of green veins remaining.

**3 Custard apple** – chlorotic interveinal areas range in colour from pale green to almost white when severe (right). The strongly bleached tissue can develop necrotic sunburnt patches between the veins.

**4 Macadamia** – chlorosis of a new growth flush where only the midribs and main lateral veins of the affected leaves are green.

5

6

**5 Macadamia** – chlorotic young leaves against a background of normal dark green mature leaves.

**6 Avocado** – young upper leaves are the first to develop chlorosis. Often the lower leaves on a chlorotic shoot will remain dark green because iron is immobile and does not move from older leaves to meet the needs of new growth flushes.

**7 Pineapple** – the young leaves are pale green to yellow often with blotches of green tissue scattered along the leaf blades (R. Broadley). In many crops iron deficiency is associated with alkaline soils or over-liming, but in pineapples symptoms are commonly a response to excess manganese found in highly acid basaltic soils. Nematodes and wet soils can also lead to iron deficiency in this crop.

7

# Manganese

Manganese availability is reduced in high pH calcareous soils but is often very high in the acid soils commonly chosen for tropical fruit and nut production. Therefore, deficiencies of manganese are less common amongst this group of crops than some other trace element deficiencies. Toxicity or nutrient imbalances caused by very high levels of available soil manganese, for example manganese-induced iron deficiency of pineapples, are far more common. However, over-liming of soils, particularly well drained, poor, coastal sandy soils, can induce deficiency.

**Function** Manganese is necessary for chlorophyll formation for photosynthesis, respiration, nitrate assimilation, and for the activity of several enzymes. The concentration of manganese in leaves can range very widely from 10 to 15 ppm, when deficient, to many thousands of parts per million when it is toxic. Manganese is only moderately mobile in plant tissues so symptoms appear on younger leaves first, most often in those leaves just reaching their full size.

**Symptoms** Manganese deficiency causes a light green mottle between the main veins. A band of darker green is left bordering the main veins (compare with iron deficiency) while the interveinal chlorotic areas become a pale green or dull yellowish colour. Symptoms are usually seen on recently matured leaves, rather than the very young or old leaves (compare with magnesium deficiency). Leaf shape and size, and the shoot length are usually normal (compare with zinc deficiency).

The first indications of deficiency in avocado is an interveinal chlorosis seen near the midrib. This symptom then spreads towards the leaf margins. Affected leaves are dull in appearance. Manganese deficiency in persimmon can reduce yields and cause defoliation, fruit drop and, in extreme cases, shoot dieback. The disorder may be recognised by the appearance of black necrotic spots in the chlorotic interveinal areas of basal leaves on new shoots.

**Corrective measures** Do not over-lime the soil but try to maintain a slightly acid pH of about 6.0. Foliar sprays are the best corrective treatment. An annual spray in spring, as soon as there is a good cover of new leaves, is usually sufficient for maintenance, but some crops may need two or three sprays per season. Apply 100 g manganese sulphate plus 100 g urea/100 L in spring or early summer when leaves of a main flush are two-thirds expanded. Soil applications can be ineffective due to immobilisation, especially in heavier soils or soils which have been over-limed.

1

3

2

4

**1 Custard apple** – a mottled pale green interveinal chlorosis develops in young and recently matured leaves. Leaf Mn = 8 ppm.

**2 Kiwifruit** – with increasing severity of the deficiency (centre right to centre left, to top leaf), the chlorotic areas becomes more yellow and expand, leaving less green tissue along the lateral and main veins. Leaf Mn = 8 ppm. The bottom leaf is healthy.

**3 Kiwifruit** – mild deficiency is first seen as a pale green mottled interveinal chlorosis spreading towards the midribs of recently matured leaves. Leaf Mn = 10 ppm.

**4 Passionfruit** – mottled yellowing between the veins of recently matured leaves. Leaf Mn = 8 ppm.

Deficiencies 55

**5 Avocado** – pale green interveinal chlorosis of a young mature leaf. Leaf Mn = 22 ppm.

**6 Persimmon** – chlorosis between the veins. Left leaf is viewed from its underside. Leaf Mn = 11 ppm.

# Zinc

Zinc deficiency is perhaps the most widespread and potentially the most growth and yield-limiting trace element deficiency of fruit crops. It commonly affects bananas, macadamias and custard apples in all the main districts and on most orchard soils, including some quite acid basaltic red loam soils (kraznozems). Problems often appear in spring when crops are growing quickly but have difficulty absorbing sufficient zinc from cold soil.

**Function** Zinc is important for the formation and activity of chlorophyll and in the functioning of several enzymes and the growth hormone auxin. The severe stunting of leaves and shoots which is so typical of zinc-deficient crops is a consequence of low auxin levels in tissue.

**Symptoms** Young leaves are usually the most affected and are small, narrow, chlorotic and often rosetted due to a failure of the shoot to elongate. In bananas, each successive leaf of the flush is smaller than the previous one, and emerges with a reddish pink coloration on its underside. The opened leaf usually loses this pink colour, but chlorotic bands develop parallel with the lateral veins and alternate with green strips, producing a 'rainbow leaf' pattern. The first indication of zinc deficiency in mango is small, slightly curved, narrow, stiff, thickened leaves in rosetted terminal flushes on the upper part of the tree. Eventually all flushes are affected. In severe deficiency, flushing may stop and twigs or even whole branches die back. Bloom spikes are small, deformed and drooping.

In young pineapple plants, zinc deficiency is indicated by the young heart leaves bunching together and then tilting horizontally. This condition is commonly called 'crook-neck'. Older plants may develop yellow spots and dashes near the margins of older leaves, that eventually coalesce into brown blister-like blemishes, giving the leaf surface an uneven feel.

**Corrective measures** Foliar sprays of 100 g zinc sulphate heptahydrate/100 L water with wetting agent will give a quick response in most crops including custard apple, pawpaw, macadamia and avocado. Bananas and pineapples can be sprayed with up to 500 g zinc sulphate/100 L. From two to four sprays applied through the main active growth period may be needed. Sprays are most effective when applied to new growth flushes (leaves are half to two-thirds fully expanded) except for custard apples where the risk of leaf burn will be minimised if sprays are delayed till the leaves are fully mature.

Soil applications of zinc usually last from two to five years but take longer to correct a deficiency, especially in heavier soils. An application of 25 g of zinc sulphate for each square metre of under-tree canopy area, spread in a 30 cm band around the drip line of the tree or vine is suitable for most crops including custard apple, lychee, avocado and passionfruit. Use only 10 g/m$^2$ for macadamias. Alternatively, a general application of 30 kg/ha of zinc sulphate (23% Zn) or 8.5 kg/ha of zinc oxide (80% Zn) has been found to be an effective treatment of the soil prior to planting and also in established plantings of bananas and other crops.

**1 Banana** – a patch of stunted, chlorotic zinc-deficient plants in the foreground. Leaf Zn = 11 ppm.

**2 Banana** – young leaves are chlorotic. The chlorosis is usually banded, with chlorotic and green bands alternating along the lamina, a condition described as 'rainbow leaf'. Youngest unfurled leaves often develop a red pigmentation on the undersides which may fade as the leaf unfurls. (F. Chalker) Leaf Zn = 9 ppm.

**3 Banana** – a closer view of the chlorotic bands found in an affected 'rainbow leaf'. (D. Stevenson)

**4 Macadamia** – a strongly patterned interveinal chlorosis which initially develops away from the leaf margin (cf. magnesium deficiency). Leaf Zn = 8 ppm.

5

7

6

8

9

**5 Macadamia** – a cluster of stunted leaves are produced at the ends of shoots ('rosetting'). This is caused through the failure of stems to elongate normally at the ends of deficient shoots. Leaf Zn = 6 ppm.

**6 Custard apple** – young leaves are small, narrow, slightly distorted and develop interveinal chlorosis.

**7 Custard apple** – chlorosis is strongly banded and tends to be more developed at the distal end of leaves. Leaf Zn = 7 ppm.

**8 Custard apple** – stunted, chlorotic 'little leaves' clustered at the end of a shoot. Normal-sized healthy leaf at right.

**9 Passionfruit** – bright yellow-green interveinal chlorosis of young leaves. Leaf Zn = 16 ppm.

**10 Passionfruit** – affected leaves are very small and narrow compared to the normal leaf below.

**11 Passionfruit** – strongly chlorotic stunted young leaves compared to healthy green normal-sized leaf left. Leaf Zn = 7 ppm.

**12 Avocado** – the youngest leaves at the ends of shoots are chlorotic, very small and often distorted or curled. Growth is clustered at the shoot tip due to a failure of this part of the stem to elongate. Leaf Zn = 14 ppm.

**13 Avocado** – milder symptoms consist of interveinal chlorosis of a few terminals where leaf size is only slightly reduced and leaf distortion is minimal. (R. Fitzell)

60     Common nutritional problems and their correction

14

15

16

Deficiencies 61

17

18 b

18 a

**14 Avocado** – a zinc-deficient tree showing very stunted, chlorotic leaves (top left) compared to normal leaves (right). (R. Broadley, QDPI)

**15 Avocado** – small round fruit are produced on zinc-deficient trees. Normal fruit at right. (R. Broadley, QDPI)

**16 Persimmon** – strongly patterned yellow-green interveinal chlorosis. Leaf Zn = 10 ppm.

**17 Pineapple** – young plants often develop the 'crook-neck' symptom, where heart leaves are hard and brittle, gather together and bend sideways. (R. Broadley, QDPI)

**18 Pineapple** – in older plants, chlorosis first develops as broken streaks of yellow (18a). The chlorotic spots extend and form light brown blisters covering much of the leaf (18b). (R. Broadley, QDPI)

# COPPER

Cases of proven copper deficiency are not common in most tropical fruit and nut crops of eastern Australia except for pineapples grown on leached coastal sandy soils in southern Queensland and occasionally in macadamia nuts in northern New South Wales.

Coastal sandy heath soils, soils derived from calcareous dune sands or marine limestone are often low in copper. Deficiency can also result from the formation of organic copper complexes in soils rich in organic matter, such as reclaimed peaty soils. Regular copper sprays applied as fungicidal treatments to some crops will usually prevent deficiency even where copper was originally low.

**Function** Copper is essential for photosynthesis, for the functioning of several enzymes, in seed development, and for the production of lignin which gives physical strength to shoots and stems.

**Symptoms** The symptoms of copper deficiency are somewhat varied, depending on the particular crop, and they are often not specific to copper deficiency. Terminal growth is often restricted, leading to dieback of twigs, death of growing points and sometimes rosetting, and multiple buds may form at the ends of twigs. Foliage can be chlorotic, uniformly pale in the case of bananas, or darker than normal or dull and brownish in colour, for example avocado.

The limbs of deficient macadamia nut trees become twisted, a symptom akin to the S-shaped branches observed in copper-deficient citrus. This twisting occurs because the new deficient shoots are weak, become bent down by their own weight, and then attempt to grow upward. In coconut palms, the rachis of the youngest leaf bends and the tip turns yellow and dries out.

Pineapples on coastal sandy soils of Queensland are often affected by copper deficiency. Growth is severely stunted and leaves are narrow, U-shaped in section, and curved downward at their tips. Tip death occurs in some young leaves.

**Corrective measures** Copper deficiency will be prevented or corrected, by the regular use of copper fungicides such as Bordeaux mixture (500 g copper sulphate (bluestone) plus 500 g fresh hydrated lime in 100 L) or copper oxychloride. Foliar application of copper, in any form, can burn the foliage of pineapples. For this crop, especially in newly cleared, heavily leached sandy soils, apply a preventative soil application of 30 kg/ha of copper sulphate. For other fruit and nut crops a soil application of 20–30 kg/ha of copper sulphate will correct the deficiency for several years.

Deficiencies 63

1a

1b

2a

2b

**1a and 1b Macadamia** – twisted limbs in a young tree (1a) and in a branch of a mature tree (1b). New shoots are weak and bent downward by the weight of the shoot. They then twist in an attempt to grow upward.

**2a Macadamia** – multiple budding in a shoot (top centre) and tip dieback commencing (top right).

**2b Macadamia** – twigs with multiple buds send out numerous soft shoots which die back, producing a 'witches broom' effect.

## 64 Common nutritional problems and their correction

**3 Pineapple** – growth is stunted and the leaves are light green and have an oily appearance. (R. Broadley, QDPI)

**4 Pineapple** – the leaves are U-shaped in section and bend downwards. (R. Broadley, QDPI)

# Boron

Boron deficiency most often affects pawpaw, custard apple, avocado and lychee. Low boron leaf analyses are commonly encountered in other crops but they less frequently develop deficiency symptoms or have their yield or growth affected. The incidence of boron deficiency is influenced by soil and climatic factors, and by cultural practices. Soils vary in their capacity to supply and retain boron. Boron is more readily leached than other trace elements especially from acid sandy soils or the highly weathered red basaltic soils.

Heavy leaching rainfall can remove much of the available boron from the surface soil while either drought or cool spring or autumn soil temperatures can reduce the root's access to and capacity to absorb soil boron. Since fruit crops require a small but continuous supply from the soil for growth, pollination and fruit development, a deficiency of boron can be induced by fluctuating seasonal conditions. This is why symptoms may suddenly appear in one season though the problem has not been seen for a number of years.

Cultural practices including over-liming, excessive irrigation, heavy application of fertilisers, particularly of potassium or nitrates, and practices which lead to root damage or soil compaction can all induce a deficiency of boron. High levels of exchangeable aluminium can also reduce boron uptake. The red kraznozem soils, often favoured for fruit growing, are naturally high in aluminium and the problem is accentuated by further acidification and organic matter depletion. Loss of soil organic matter under orchard culture also reduces the capacity of soils to retain boron, leading to an increased incidence of deficiency in older orchards.

**Function** Boron is important for cell division and development in the growth regions of the plant near the tips of shoots and roots. It also affects sugar transport and appears to be associated with some of the functions of calcium. Boron affects pollination and the development of viable seeds which, in turn, affect the normal development of fruit. A shortage of boron also causes cracking and distorted growth in fruit.

Boron does not easily move around the plant and, therefore, the effects of deficiency appear first, and are usually most acute, in young tissues, growing points, root tips, young leaves and developing fruit.

**Symptoms** Symptoms may affect fruit, shoots, and leaf growth but unless the deficiency is severe the symptoms in the fruit are usually noticed first.

The fruit of boron-deficient pawpaw are deformed and bumpy due to the irregular fertilisation and development of seeds within the fruit. Ripening is uneven and the developing fruit secrete pinkish white to brown latex. Heavy premature shedding of deficient male tree flowers and impaired pollen tube development can lead to poor set in the fruit-bearing female trees. Upper mature leaves are pale, stiff and brittle, and may die at the tip, curling downward to become claw-like in shape. Growth ceases at the growing point and a white exudate flows from cracks in the upper trunk which becomes twisted. The growing point and stubby terminal roots may die.

In pineapple, fruitlets separate and the gaps between them become corky. Root growth is poor but only when the deficiency is severe does the growing point die or the leaves show symptoms. Deficient avocado trees produce small misshapen fruit, usually twisted to one side giving them a dumpy, lop-sided and bulbous shape. An indented blemish sometimes appears on the concave side of these fruit. Leaf symptoms are not usually seen. The variety Sharwil is particularly sensitive and is a good indicator of low boron in the orchard.

**Corrective measures** Except for avocados and macadamias, foliar sprays usually give a quicker response and can be safer in established crops than soil dressings. Several sprays may be needed (for example, monthly during the main growing season for pawpaw). Boron sprays are not recommended for avocados, as they can burn flowers and leaves.

**Foliar spray**
150 g/100 L polyborate powder (20.5% B).

**Soil application**
The rate for a general application over the whole orchard area can range from 10–40 kg/ha borax (10.5% B) with 20–30 kg suiting most crops and soils including bananas and pineapples. The rate for pawpaws should not exceed 20 kg/ha. Alternatively an under-tree application of borax at 2 g/m$^2$ of soil area covered by the tree or vine canopy is suitable for avocado, lychee, macadamia, passionfruit, and guava. For deficient custard apples use 5 g/m$^2$ up to twice per season.

Excess boron is highly toxic to plants. When applying borax to the soil ensure that it is spread evenly over the area to be treated and that all lumps are crushed to avoid 'hot spots'. Alternatively, polyborate fertiliser (at half the borax rate) can be dissolved in water and sprayed onto the soil.

Deficiencies 67

**1 Pawpaw** – bumpy fruit. Many flowers shed and few fruit set. Those which develop are deformed and often a white latex exudes from the skin during fruit growth. Most of the seeds are poorly developed or aborted.

**2 Pawpaw** – severe deficiency causes stunting, distortion and necrosis of upper leaves, latex exudation from the stem, and finally death of the growing point. (K.R. Chapman)

**3 Pawpaw** – severe stunting and distortion of young leaves. Necrosis and growth restriction around the margin causes severe cupping and distortion of the leaves. (K.R. Chapman)

**4 Avocado** – low boron supplies at flowering and fruit development stages can lead to the fruit aborting or developing abnormal shapes. Most commonly the fruit is bulbous and twisted to one side. (R. Fitzell)

**5 Avocado** – narrow distorted young leaves with necrotic and chlorotic margins and tips. In severe cases of the deficiency, growing points may die. (R. Fitzell)

**6 Pineapple** – cracking and cork formation on and between fruitlets. (R. Broadley, QDPI)

# MOLYBDENUM

Although molybdenum deficiency occurs widely in many soils and pasture legumes, vegetable crops and occasionally cereals are affected, there has been no confirmed case of molybdenum deficiency in field-grown fruit-tree or nut crops in Australia. The only report of molybdenum deficiency affecting fruit crops in the field is the condition described as 'yellow spot' in citrus from Florida, USA. Symptoms consist of large yellow spots on the upper surface of the leaf with gum and corky cells forming on the lower side.

Although the fruit crop may not respond directly to molybdenum, the use of molybdenum fertilisers are often essential to establish and maintain a successful legume-based cover crop in the orchard. Molybdenised single superphosphate applied at 250–500 kg/ha is the usual means of satisfying this need for 2–5 years.

# Toxicities

Some elements, particularly nitrogen, chloride, sodium, boron and manganese, can cause injury when taken up in excess.

## SODIUM AND CHLORIDE

Most fruit crops are sensitive to high concentrations of salt (sodium chloride) occurring in the soil or in irrigation water. The heavy leaching rains and well drained soils of most tropical fruit growing areas usually ensures that cases of salt toxicity are infrequent compared to the incidence in temperate fruit crops of our inland irrigation areas. However, isolated cases of salt toxicity have been found in areas adjacent to coastal tidal rivers where soils are salt-affected or irrigation water becomes brackish.

Absorption of salt deposited on leaves from onshore winds is another cause of salt injury to trees in exposed locations close to the sea. This condition has been identified in kiwifruit, avocado and guava in coastal New South Wales.

Burning of the leaf tip and yellowing or scorching of the margins are the most common symptoms of chloride or sodium toxicity. Leaf fall can be heavy and dieback may follow. Older leaves usually show the symptoms first. Similar symptoms can be caused by drought, boron toxicity, potassium deficiency or fertiliser burn, so leaf analysis may be needed to confirm the diagnosis. If affected leaves contain more than 0.5% chloride or 0.2% sodium, the injury could be related to excess salt. While an excess of either sodium or chloride can cause toxicity, high leaf-chloride is more common. This is because sodium is less freely translocated to the leaves of fruit trees.

Irrigation management and drainage to minimise salt build-up in the root zone are important means of combatting salinity damage to crops. Where a problem exists, have the irrigation water and soil tested to determine the source of salinity. Use of saline water (E.C. > 3 dS/m) with overhead irrigation may increase salt damage through chloride uptake by leaves. This uptake will be lessened by watering at night, when evaporation is lower, or by using under-tree irrigation systems to avoid wetting the leaves.

**Cl1 Guava** – light brown burn of the margin of mature leaves. A light green chlorosis precedes the development of the necrosis. Leaf analysis Cl = 2.9%, Na = 0.37%.

**Cl2 Kiwifruit** – light grey to brown burn around the leaf margin commencing at the tip and quickly surrounding the whole lamina. The green central part of the leaf tends to be a dark bluish green with a transitional grey-green border separating the green and necrotic tissues. Leaf analysis Cl = 2.9%, Na = <0.1%.

**Cl3 Avocado** – irregular light brown burn especially near the tip. Leaf Cl = 0.96%.

4

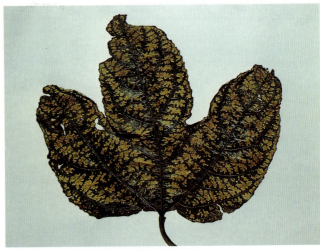

5

**Cl4 Avocado** – chlorosis usually precedes the necrosis. Leaf Cl = 0.85%.

**Cl5 Passionfruit** – light brown necrosis of the interveinal areas commencing at the leaf margin and progressing inward towards the main veins. Leaf Cl = 3.1%.

**Cl6 Passionfruit** – a salinity-induced nutrient imbalance of high chloride and sodium, with very low potassium uptake, caused this burn of the leaf tips and margins. Leaf analysis Cl = 1.9%, Na = 0.42% and K = 0.45%.

6

# BORON

**B**oron toxicity often affects fruit crops grown in inland areas of Australia. The main causes are using irrigation water which is high in boron, or poor drainage, or irrigation practices which concentrate soluble boron into the root zone. However, boron toxicity is rarely a natural occurrence in tropical fruit crops which are grown in high rainfall districts. Most often it is the result of excessive or uneven application of boron fertilisers to the soil or in foliar sprays. Great care should be taken to apply boron fertilisers according to recommendation, as the range between deficiency and toxicity is quite narrow for this trace element.

The symptoms usually commence as marginal and/or interveinal chlorosis surrounding brown necrotic spots. These spots tend to coalesce into extensive burnt areas of tissue. Older leaves are affected first, but the symptoms spread progressively to younger leaves.

**1 Custard apple** – interveinal chlorosis in young leaves and brown necrotic spots of the interveinal tissue of mature and older leaves in a glasshouse-grown plant. Leaf B = 270 ppm (upper leaves) and 430 ppm (lower leaves).

**2 Custard apple** – range of symptoms (left to right) in young and mature to old leaves of glasshouse-grown plants, showing interveinal chlorosis to spotting of interveinal and marginal leaf tissue.

**3 Custard apple** – toxic symptoms in a field-grown tree, consisting of marginal and interveinal yellowing with brown marginal necrosis of older leaves. (G. Sanewski, QDPI)

**4 Macadamia** – creamy yellow chlorosis of the margin and interveinal tissue which may become necrotic if toxicity is severe. (D. Batten)

**5 Pawpaw** – severe interveinal scorch of older leaves, with cupping, marginal necrosis and chlorosis of upper leaves. (K.R. Chapman)

# MANGANESE

Manganese toxicity is almost always associated with acid soils and the severity is made worse by periodic waterlogging or use of acidifying ammonium fertilisers, especially ammonium sulphate (see Table 7). Manganese toxicity occurs on a range of soil types, but the red basaltic soils of coastal northern New South Wales and southern Queensland are particularly high in available manganese.

In macadamia, manganese toxicity causes an orange-yellow chlorosis around the margin at the distal ends of older leaves. Within these chlorotic areas, reddish brown 1–2 mm spots appear. The toxicity often occurs in combination with magnesium deficiency. Bananas usually show no obvious symptoms and their growth is rarely stunted, even though manganese commonly accumulates to very high concentrations (exceeding 5000 ppm) in the leaves. High manganese uptake reduces the uptake of calcium, magnesium and zinc and shortens the green life of harvested fruit causing the fruit to arrive at market in a 'mixed ripe' condition (Turner 1985*).

'Green blotch', a disorder of persimmon fruit, is associated with high uptake of manganese in fruit, which is also reflected in the leaf analysis. The symptom appears as a green discoloration of the skin particularly around the fruit apex. This region is often depressed and can turn a brownish black if the condition is severe.

Control measures are usually directed towards correcting soil acidity but drainage should be examined in affected patches and, if necessary, corrected. Overcoming soil acidity in a producing orchard is a slow process, particularly acidity at depth. Applications of lime or preferably dolomite (if magnesium is low) at 500 kg to 1 tonne/ha, repeated each year, may take up to 4 years to correct a manganese toxicity problem.

* Turner, D.W. (1985) – Fertilising bananas: calcium, magnesium and trace elements. Agfact H6.36 Department of Agriculture, NSW.

1 **Macadamia** – orange-yellow chlorosis of the distal margins of older leaves. Small reddish brown spots of manganese oxide form within the chlorotic areas. Note the dull pale green chlorosis (right leaf) which is evident prior to the golden chlorosis, and the faint brownish spots already forming. Leaf Mn = 4700 ppm and leaf Mg = 0.08% (satisfactory).

**3 Custard apple** – an irregular chlorotic pattern develops around the outer parts of older leaves in this glasshouse-grown seedling. Small brown specks of manganese oxide are scattered within the greener areas. Leaf Mn = 2800 ppm.

**4 Custard apple** – field symptom. Small brown necrotic spotting (left), with more severe interveinal necrosis (right) and the breaking away of dead marginal and interveinal tissue. (G. Sanewski, QDPI)

**5 Avocado** – interveinal chlorosis and tiny reddish brown spots form in the interveinal areas of older leaves. Finally the leaf dies from the tip. Leaf Mn = 942 ppm.

**6 Avocado** – small brown specks of manganese oxide deposit along the veins at an early stage of symptom development while interveinal chlorosis is still only faintly evident. Leaf Mn = 4060 ppm.

**7 Persimmon** – tiny brown spots on the skin of fruit from a tree affected by manganese toxicity. (J. Campbell, QDPI)

**8 Persimmon** – leaf symptom of combined magnesium deficiency (interveinal chlorosis) and manganese toxicity (small brown manganese specks within the chlorotic interveinal areas and bordering the lateral veins). (J. Campbell)

**2 Macadamia** – combined magnesium deficiency and manganese toxicity. High soil manganese and low magnesium occur together in many acid soils. Compare with the previous photograph and with the photograph of magnesium deficiency and note the small brown spots associated with manganese accumulation which distinguish this condition from normal magnesium deficiency. Leaf Mn = 5450 ppm and leaf Mg = 0.02% (deficient).

**9 Passionfruit** – light green interveinal chlorosis of a mature juvenile leaf showing brownish specks in the interveinal areas and adjacent to lateral veins. Leaf Mn = 2200 ppm.

**10 Kiwifruit** – interveinal chlorosis and dark brown spots adjacent to the veins is followed by death of marginal and interveinal tissue. Leaf Mn = 1160 ppm.

# NITROGEN (FERTILISER BURN)

Excessive or careless application of any nutrient may cause injury, but symptoms seen soon after the application of fertiliser are mostly caused by the sudden uptake of soluble nitrogen salts. Young trees and plants in nursery pots are particularly at risk if fertiliser application is heavy-handed, spread unevenly or concentrated near the tree butt. Putting soluble fertiliser into the planting hole can cause serious root injury often leading to tree death. Pineapple leaves can also be burned and the growing point damaged from direct contact with solid fertiliser during side-dressing.

Symptoms of fertiliser burn develop suddenly, within a few days of application, except when fertiliser is not watered in, when injury may not be seen until after the first shower of rain. Leaf curl, defoliation, dieback and fruit drop are the most obvious effects. Sometimes, young leaves develop translucent dark green damaged areas which later become brown and dead. The lower leaves of young trees are usually shed first. Abscission may take place between the leaf blade and the petiole, leaving petioles attached to the stem. Young shoots die at the tip. Although difficult to see, root injury appears as a brown discoloration of rootlets. If direct contact has occurred between roots and fertiliser granules, as might occur in planting hole applications, dark brown scars may be seen at contact points.

1

**1 Pineapple** – damage to leaves caused by direct contact of fertiliser with leaves during broadcasting operations. (R. Broadley, QDPI)

# Some non-nutritional symptoms

*A* number of non-nutritional stresses can cause symptoms in leaves, fruit or other tissues which can be mistaken for nutrient disorders. However, the resemblance is often only superficial and basic differences in the patterns can be seen when the specimen is examined closely.

Non-nutritional symptoms can arise from several environmental stresses – such as frost, heat, drought or wind – through physiological effects resulting from root injury or waterlogging, or from spray burns, herbicide injury, genetic abnormalities, fungal attack, virus infection, or insect injury.

**Vein chlorosis** The terms 'vein chlorosis', 'vein clearing' or 'yellow vein' are used to describe the loss of green colour from the midrib and major veins. The interveinal tissue remains green, though it may become slightly paler than normal. This pattern is contrary to common nutritional patterns where colour loss begins between the veins.

Vein chlorosis often indicates damage to the roots or vascular system as a result of girdling of the trunk or prolonged waterlogging. Some herbicides, including bromacil and diuron, can produce vein chlorosis symptoms in affected trees.

**Herbicides** Herbicides cause a variety of symptoms in leaves when taken up either through the roots or by direct contact of the sprays with leaves, branches or the tree butt. Young trees are particularly vulnerable to incorrect application rates or methods, or from spray drift.

Apart from possible tree death, herbicides cause stunting, distortion of leaves or shoots, vein clearing, and a variety of chlorotic leaf patterns. Chlorosis caused by herbicides usually differs from typical nutrient deficiency symptoms, in that pattern symmetry is usually absent and the chlorotic areas are unrelated to leaf venation. The colour of the chlorosis is often an artificially bright hue of yellow, orange, cream or white.

**Spray burn** Sprays of trace elements and other nutrients or pesticides can injure leaves, flowers or fruit, if they are applied incorrectly, for example at the wrong strength or time, during heatwave conditions, to plants that are drought-stressed, or to immature growth. Plants which have been 'softly' grown in a glasshouse or shadehouse can be injured by spray concentrations normally safe for field-hardened plants. Injury can occur if materials are not correctly mixed or not kept agitated in the spray vat. Where several materials are used together in the one spray, ensure they are compatible, that is, they won't react to produce a residue that could damage leaves or fruit.

Injury symptoms of spray burn will vary with the spray materials, the conditions at spraying, the spray equipment and the type and condition of the plant and the maturity of the tissue at spraying time.

A clue to this type of damage can often be observed in the injury pattern, which frequently follows the shape and distribution of the spray droplets. Other indications are the suddenness with which injury occurs after application and that many plants or trees develop the symptoms simultaneously. Nutrient disorders tend to show up to different degrees within the tree and throughout the orchard; the symptoms may appear first in only a few scattered trees, then gradually in others over several weeks or months.

Symptoms caused by adverse environmental factors such as frost, hail, heat or wind, also tend to appear within hours or at most a few days of some sudden weather event and this can often help identify the cause.

**Frost and cold injury** Many tropical fruit crops, including custard apples and pineapples, are very sensitive to frost or a period of low temperatures, particularly at flowering time, when a new flush of growth is developing or when fruit is on the trees. Young trees are especially sensitive to low temperatures. Injury will usually be more severe in low lying areas of the orchard and places where air drainage is impeded due to large trees or other barriers. Sections of more exposed leaves or other tissues develop scorched areas. Affected leaves turn yellow or brown or may die, while the more exposed parts of fruit skins will discolour, russet or split. Frosted fruit may subsequently grow into grotesque shapes or become susceptible to entry of fungal or bacterial rots. In severe cases the growing point or crown of the plant is killed.

**Sunburn, heat stress and drought** When leaves curl because of moisture shortage or very hot winds, the undersides of leaves are exposed to summer sunlight and may develop grey, brown or reddish sunburnt areas. Severe moisture stress can cause irregular-shaped burning of the outer parts of leaves which may then yellow and die. Sunscald, caused by sudden heatwave conditions can produce similar effects on leaves or fruit even when soil moisture is adequate. The distribution of healthy and damaged tissue is often related to the presence or absence of cover from protecting leaves.

Pineapples are often subject to heavy fruit loss during very hot days especially in the ratoon crop. Mild injury causes a yellowing on the exposed side of the fruit, but severe burning results in sunken brown skin lesions in the centre of the yellowed skin. Fruit tissue under the severely damaged skin is susceptible to breakdown. Exposed fruit of custard apples are also subject to sunburn which causes blackened and yellowed areas on the side facing midday sun.

**Hail damage** Hail can damage both leaves and fruit. A severe storm may leave a tree defoliated, while minor damage includes torn or perforated leaves. Fruit nearing maturity are most vulnerable to injury and may subsequently develop fungal or bacterial rots.

**Wind damage** Strong winds distort new growth and tatter the leaves. Fruit skins become scarred by the rubbing of leaves and twigs. New plantings without windbreaks are most susceptible.

**Genetic abnormalities** These tend to be found as a random scatter of affected plants throughout a crop, rather than in patches as usually happens with nutritional disorders. Isolated plants or just a branch

# Some non-nutritional symptoms

**1 Bananas** – injury from the herbicide Roundup causes narrow yellow leaves in bananas. The leaf margins are straight and parallel giving the leaf a rectangular shape (upper right leaf of foreground plant).

**2 Macadamia** – new growth on macadamia trees affected by Roundup spray drift emerges as chlorotic young leaves with a proliferation of multiple buds and stunted distorted shoots. (R. Fitzell)

**3 Avocado** – severely distorted growth characterises hormone spray damage. (R. Fitzell)

**4 Avocado** – chlorotic and necrotic patches in these leaves followed tree injection with phosphorus acid to control root injury by the fungus *Phytophthora*.

within a tree may be stunted, distorted or show irregular yellow patterns in leaves, usually unrelated to leaf venation. Fruit can also be affected by abnormal shapes or colours.

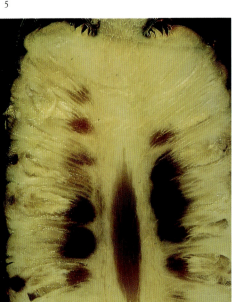

**5 Custard apple** – cracking can result when the soil moisture supply is irregular during fruit development. Heavy application of nitrogen fertiliser in autumn can also cause splitting.

**6 Passionfruit** – collapse of a section of a tender young shoot and leaves after a heavy frost. (R. Fitzell)

**7 Custard apple** – frost or severe cold weather can cause areas of the fruit skin to blacken.

Some non-nutritional symptoms 83

10

11

12

13

8  **Pineapple** – 'blackheart', a browning of the flesh near the core, commonly occurs in fruit maturing in cool cloudy weather or in harvested fruit stored under refrigeration. (M. Stevenson, QDPI)

9  **Banana** – stunted distorted leaves bunched at the top of a plant affected with the 'bunchy top' virus. (D. Stevenson)

10  **Macadamia** – sudden hot weather in spring can cause yellowing of new growth which may be confused with iron deficiency symptoms.

11  **Banana** – streaky chlorosis of a leaf affected by a virus.

12  **Avocado** – chlorosis of lateral veins and minor veinlets in a leaf affected by the sunblotch virus. Vein chlorosis is rarely the result of a nutritional disorder.

13  **Passionfruit** – irregular pattern of chlorosis and leaf distortion caused by the 'bullet' virus. (R. Fitzell)

## 84 Common nutritional problems and their correction

14

16

15

17

**14 Passionfruit** – leaf spots resulting from infection by the fungus *Alternaria*.

**15 Avocado** – genetic abnormalities often cause unusual shapes or colours in fruit or leaves. This chimaera has caused irregularly spaced light green bands alternating with normal skin pigmentation. (R. Fitzell)

**16 Macadamia** – genetic mutation produced these variegated leaves on a new shoot. Leaves on other branches are a normal green colour. Note the irregular pattern of the chlorosis, completely unrelated to the veins. This usually indicates a non-nutritional cause.

**17 Macadamia** – brown necrosis of the margins and tips of older leaves caused by the fungus *Dothiorella*. These symptoms could be confused with potassium deficiency possibly requiring specialist advice or leaf analysis to confirm the diagnosis.

# Appendix

**Tables of leaf analysis standards**  The following standards were used for diagnostic leaf analyses conducted at the NSW Department of Agriculture Chemistry Branch laboratories from 1958 to 1988. They are based on information from world literature and from local research, crop surveys and diagnostic analyses. Some principal references are listed at the end of this appendix and reference numbers of particular relevance to specific crops are given at the bottom of each table.

**Leaf sampling**  The illustrations below show the sample tissue used for plant analysis of some crops. The correct sampling procedure must be followed or the analytical results will be misleading.

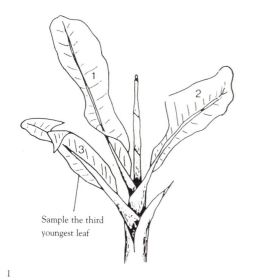

Sample the third youngest leaf

1

2

3

**1 Banana** – select the third youngest fully emerged leaf from medium-sized suckers which can be reached from the ground. Take sample strips of lamina from at least 10 plants per block.

**2 Banana** – 20 cm wide strips of leaf blade tissue are taken half way along the leaf on each side of the midrib.

**3 Custard apple** – the youngest mature (usually 4th) leaf from non-fruiting shoots.

86   Appendix

4

5

6

Appendix    87

7

8

**4 Guava** – take the third pair of fully developed leaves back from the tip of fruiting terminals.

**5 Kiwifruit** – sample the first leaf above the fruit towards the growing point.

**6 Macadamia** – choose a mature leaf from the second whorl of current season's growth, avoiding terminals carrying new flushes.

**7 Mango** – the most recently mature leaf is taken in early spring, just prior to flowering.

**8 Passionfruit** – the youngest fully expanded leaf is taken from well developed actively growing laterals.

# Avocado
## (*Persea americana*)

**Sampling time**  Early autumn i.e. April–May for tropical and subtropical localities (Northern Hemisphere October–November)

**Plant part and growth stage**  Recently matured fully expanded 4–5 month old summer flush leaves from non-fruiting terminals

| Nutrient | Deficient | Low | Normal | High | Excess |
|---|---|---|---|---|---|
| Nitrogen *Fuerte* % (N) | <1.4 (0.8$^D$–1.35$^D$) | 1.4–1.5 | 1.6–2.0 | >2.0 | |
| *Hass* | <1.4 | 1.4–2.2 | 2.2–2.6 | >2.6 | |
| Phosphorus % (P) | <0.05 | 0.05–0.07 | 0.08–0.25 | 0.26–0.30 | >0.30 |
| Potassium % (K) | <0.35 (0.22$^D$) | 0.35–0.74 | 0.75–2.0 | 2.1–3.0 | >3.0 |
| Calcium % (Ca) | <0.4 | 0.4–0.9 | 1.0–3.0 | 3.1–4.0 | >4.0 |
| Magnesium % (Mg) | <0.15 | 0.15–0.24 | 0.25–0.80 | 0.9–1.0 | >1.0 |
| Sulphur % (S) | <0.05 | 0.05–0.19 | 0.2–0.3 | 0.4–1.0 | >1.0 |
| Sodium % (Na) | | | 0.01–0.04 | 0.10–0.20 | >0.25 |
| Chloride % (Cl) | | | <0.25 | 0.25–0.5 | >0.5 (1.4$^T$–1.9$^T$) |
| Copper[1] ppm (Cu) | <3 | 3–4 | 5–15 | 25–300 | |
| Zinc[1] ppm (Zn) | <15 (12$^D$–14$^D$) | 15–29 | 30–50 | 60–300 | |
| Manganese[1] ppm (Mn) | <15 | 15–29 | 30–500 | 550–1000 | >1200 (1580$^T$–2614$^T$) |
| Boron ppm (B) | <10 (6$^D$–8$^D$) | 10–20 | 30–100 | 150–180 | >200 |

[1] Values for copper, zinc or manganese in leaves sprayed with fungicides or nutrient sprays containing trace elements cannot give a reliable guide to nutritional status even when leaves are washed.

$^D$ Level found in field specimens showing deficiency symptoms.

$^T$ Level found in field specimens showing toxicity symptoms.

Main information sources and further reading: 5, 9, 12

# Banana
### (Musa spp.)

**Sampling time**   During active growth

**Plant part and growth stage**   Take 20 cm strips of lamina from both sides of the midrib midway along the third youngest leaf of medium-sized suckers

| Nutrient | Deficient | Low | Normal | High | Excess |
|---|---|---|---|---|---|
| Nitrogen % (N) | <2.6 (1.9$^D$–2.3$^D$) | 2.6–2.9 | 3.0–4.0 | >4.0 | |
| Phosphorus % (P) | <0.13 | 0.13–0.18 | 0.19–0.25 | >0.25 | |
| Potassium % (K) | <2.4 (0.4$^D$–1.9$^D$) | 2.4–2.9 | 3.0–4.0 | >4.0 | |
| Calcium % (Ca) | <0.36 | 0.36–0.69 | 0.70–1.25 | >1.25 | |
| Magnesium % (Mg) | <0.20 (0.03$^D$–0.19$^D$) | 0.20–0.29 | 0.30–0.45 | >0.46 | |
| Sulphur % (S) | <0.10 | 0.10–0.19 | 0.20–0.27 | >0.30 | |
| Sodium % (Na) | | | 0.01–0.02 | 0.15–0.26 | |
| Chloride % (Cl) | | | 0.5–1.1 | 1.2–1.6 | >1.6 (2.2$^T$–3.1$^T$) |
| Copper ppm (Cu) | <3 | 3–5 | 6–20 | | |
| Zinc ppm (Zn) | <13 (7$^D$–11$^D$) | 13–19 | 20–35 | >35 | |
| Manganese ppm (Mn) | <10 | 10–29 | 30–2200 | 4000–6000 | >6000 (9435$^T$–13 100$^T$) |
| Iron[1] ppm (Fe) | | | 70–200 | | |
| Boron ppm (B) | <10 (4$^D$–8$^D$) | 10–13 | 14–70 | 80–100 | >300 (500$^T$) |

[1] Leaf analysis is not a reliable guide to iron deficiency because of surface contamination with soil, or immobility of iron within the plant, or the presence of physiological inactive iron within tissues.

[D] Level found in field specimens showing deficiency symptoms.

[T] Level found in field specimens showing toxicity symptoms.

Main information sources and further reading: 9, 11, 12

# Coconut
*(Cocos nucifera)*

**Sampling time**  Varies, usually at the start of the dry season

**Plant part and growth stage**  Select 3 pairs of healthy undamaged leaflets from the central part of leaf number 14 (13 leaves behind the youngest fully expanded leaf)

| Nutrient | Deficient | Low | Normal | High | Excess |
|---|---|---|---|---|---|
| Nitrogen % (N) | | | 1.8–2.0 | | |
| Phosphorus % (P) | | | 0.12–0.14 | | |
| Potassium % (K) | | | 0.8–1.4 | | |
| Calcium % (Ca) | | | 0.3–0.5 | | |
| Magnesium % (Mg) | | | 0.20–0.35 | | |
| Sodium % (Na) | | | 0.2–0.4 | | |
| Chloride % (Cl) | <0.1[1] | 0.1–0.4 | 0.5–0.6 | | |
| Copper ppm (Cu) | | 4–5 | 6–14 | | |
| Zinc ppm (Zn) | | <15? | 16–60 | | |
| Manganese ppm (Mn) | | 20–50 | 80–200 | | |
| Boron ppm (B) | | 8–10 | 11–20 | | >200 |

[1] Coconuts are reported to respond to chloride at leaf levels below this value.

# Coffee
## (*Coffea arabica*)

**Sampling time**  Varies, February–April or September–October for Papua New Guinea and north-eastern Australia

**Plant part and growth stage**  Fourth pair of leaves from the tip of actively growing fruiting branches

| Nutrient | Deficient | Low | Normal | High | Excess |
|---|---|---|---|---|---|
| Nitrogen % (N) | <2.2 | 2.2–2.5 | 2.6–3.4 | >3.4 | |
| Phosphorus % (P) | <0.10 | 0.10–0.14 | 0.15–0.20 | >0.20 | |
| Potassium % (K) | <1.5 | 1.5–1.9 | 2.0–2.6 | >2.6 | |
| Calcium % (Ca) | <0.4 | 0.4–0.5 | 0.6–1.6 | >1.6 | |
| Magnesium % (Mg) | <0.16 | 0.16–0.24 | 0.25–0.50 | >0.50 | |
| Sulphur % (S) | <0.08 | | 0.15–0.26 | | |
| Copper ppm (Cu) | <4 | 5–9 | 10–50 | >50 | |
| Zinc ppm (Zn) | <10 | 10–14 | 15–30 | >30 | |
| Manganese ppm (Mn) | <25 | 25–35 | 40–300 | 300–700 | >700 |
| Boron ppm (B) | <20 | 20–39 | 40–150 | 160–200 | >200 |

Main information sources and further reading: 1, 9

# Custard Apple
(*Annona* spp.)

**Sampling time**  Late February to April (Northern Hemisphere August–October)

**Plant part and growth stage**  Youngest mature leaf (usually fourth from growing point), from non-fruiting shoots

| Nutrient | Deficient | Low | Normal | High | Excess |
|---|---|---|---|---|---|
| Nitrogen % (N) | 1.5$^D$–1.9$^D$ | 2.0–2.2 | 2.5–3.3 | 3.5–3.8 | >3.8 |
| Phosphorus % (P) | <0.11 | 0.12–0.15 | 0.16–0.25 | 0.28–0.35 | >0.41 |
| Potassium % (K) | 0.2$^D$–0.7$^D$ | 0.8–1.0 | 1.2–1.6 | 1.8–2.0 | >2.4 |
| Calcium % (Ca) | | 0.4–0.5 | 0.6–1.0 | 1.5–2.0 | 3.3$^I$ |
| Magnesium % (Mg) | 0.08$^D$–0.12$^D$ | 0.15–0.18 | 0.20–0.50 | 0.70 | |
| Sodium % (Na) | | | 0.01–0.08 | 0.15–0.20 | >0.20? |
| Chloride % (Cl) | | | <0.6 | 0.8 | >1.4 |
| Copper$^1$ ppm (Cu) | | <5 | 6–60 | 100–250 | |
| Zinc$^1$ ppm (Zn) | 7$^D$–10 | 12–15 | 16–40 | >100 | |
| Manganese$^1$ ppm (Mn) | | <20 | 30–300 | 350–500 | 1200$^T$–2200$^T$ |
| Boron ppm (B) | | | 20–80 | 100–170 | |

$^1$ Values for copper, zinc or manganese in leaves sprayed with fungicides or nutrient sprays containing trace elements cannot give a reliable guide to nutritional status even when leaves are washed.
$^I$ Imbalance
$^D$ Level found in field specimens showing deficiency symptoms.
$^T$ Level found in field specimens showing toxicity symptoms.
Main information sources and further reading: 1, 9, 12

# Guava
## (*Psidium gaujava*)

**Sampling time**  Mid summer i.e. January–February (Northern Hemisphere July–August)

**Plant part and growth stage**  Third pair of fully developed leaves back from the tip of fruiting terminals

| Nutrient | Deficient | Low | Normal | High | Excess |
|---|---|---|---|---|---|
| Nitrogen % (N) | 1.1$^D$ | | 1.3–1.8 | 2.1–2.7 | >3.0 |
| Phosphorus % (P) | <0.08 | 0.12–0.13 | 0.14–0.20 | 0.21–0.25 | >0.40 |
| Potassium % (K) | <0.9 (0.5$^D$–0.7$^D$) | 0.9–1.2 | 1.3–1.8 | 2.0–2.9 | >3.7 |
| Calcium % (Ca) | <0.5 | 0.6–0.7 | 0.8–1.5 | >1.8 | |
| Magnesium % (Mg) | <0.20 | 0.20–0.24 | 0.25–0.40 | | |
| Sodium % (Na) | | | 0.01–0.20 | 0.3–0.4 | >0.4 |
| Chloride % (Cl) | | | <1.2 | 1.5 | >1.8 (2.1$^T$–2.9$^T$) |
| Copper ppm (Cu) | | | 6–40 | | |
| Zinc ppm (Zn) | | 19 | 25–40 | | |
| Manganese ppm (Mn) | <30 | 30–50 | 60–400 | | |

$^D$ Level found in field specimens showing deficiency symptoms.
$^T$ Level found in field specimens showing toxicity symptoms.
Main information sources and further reading: 1, 9, 12

# Kiwifruit
(*Actinidia deliciosa*)

**Sampling time**  Late summer i.e. February (Northern Hemisphere August)

**Plant part and growth stage**  The first leaf above the fruit, towards the growing point

| Nutrient | Deficient | Low | Normal | High | Excess |
|---|---|---|---|---|---|
| Nitrogen % (N) | <1.5 ($1.0^D$–$1.3^D$) | 1.5–2.1 | 2.2–3.0 | 3.1–5.5 | >5.5 |
| Phosphorus % (P) | <0.13 | 0.13–0.17 | 0.18–0.25 | >0.25 | |
| Potassium % (K) | <0.7 ($0.21^D$–$0.63^D$) | 0.7–1.1 | 1.8–3.0 | 3.1–4.0 | >4.8 |
| Calcium % (Ca) | <0.2 | | 2.0–4.0 | | |
| Magnesium % (Mg) | <0.10 | 0.15–0.24 | 0.3–0.7 | 0.8–1.0 | 1.2–1.5[I] |
| Sulphur % (S) | | | 0.25–0.45 | | |
| Sodium % (Na) | | | <0.05 | | |
| Chloride % (Cl) | <0.2 | | 0.4–1.0 | 1.1–1.5 | 1.6–3.2 |
| Copper[1] ppm (Cu) | <3 | | 10–25 | | |
| Zinc[1] ppm (Zn) | <12 ($8^D$–$10^D$) | 12–15 | 15–28 | | >1100 |
| Manganese[1] ppm (Mn) | <30 | 30–50 | 50–400 | 400–1500 | >1500 |
| Iron[2] ppm (Fe) | <60 | | 80–200 | | |
| Boron ppm (B) | <20 | | 30–60 | | >100 |

[1] Values for copper, zinc or manganese in leaves sprayed with fungicides or nutrient sprays containing trace elements cannot give a reliable guide to nutritional status even when leaves are washed.

[2] Leaf analysis is not a reliable guide to iron status.

[I] Imbalance, from crop affected by potassium deficiency.

[D] Level found in field specimens showing deficiency symptoms.

Main information sources and further reading: 4, 9, 12

# Lychee
*(Litchi chinensis)*

**Sampling time**  Early fruit development, October–November (Northern Hemisphere April–May)

**Plant part and growth stage**  Most recently matured leaf pair behind the flower panicle. Retain the central two leaflets

| Nutrient | Deficient | Low | Normal | High | Excess |
|---|---|---|---|---|---|
| Nitrogen % (N) | <1.0 (0.90[D]) | 1.2 | 1.3–1.5 | 1.6–2.2 | >2.2 |
| Phosphorus % (P) | <0.11 | 0.11–0.13 | 0.14–0.20 | >0.24 | |
| Potassium % (K) | <0.5 (0.37[D]) | 0.5–0.6 | 0.8–1.2 | | |
| Calcium % (Ca) | | 0.3–0.5 | 0.6–1.5 | 1.8–2.5 | |
| Magnesium % (Mg) | | 0.15–0.18 | 0.20–0.50 | | |
| Sodium % (Na) | | | <0.08 | | |
| Chloride % (Cl) | | | 0.1–0.3 | | |
| Copper[1] ppm (Cu) | | <4 | 6–25 | | |
| Zinc[1] ppm (Zn) | | <12 | 15–30 | >150 | |
| Manganese[1] ppm (Mn) | | <20 | 50–300 | >400? | |
| Boron ppm (B) | | <12 | 25–60 | | |

[1] Values for copper, zinc or manganese in leaves sprayed with fungicides or nutrient sprays containing trace elements cannot give a reliable guide to nutritional status even when leaves are washed.

[D] Level found in field specimens showing deficiency symptoms.

Main information sources and further reading: 1, 9, 12

# Macadamia
(*Macadamia integrifolia* and *M. tetraphylla*)

**Sampling time** Spring, when all leaves have hardened and no new growth is evident i.e. September–December (Northern Hemisphere March–June)

**Plant part and growth stage** Mature leaf (6–7 months old) from the second whorl of current season's growth. Avoid terminals bearing new flushes >1.5 cm long

| Nutrient | Deficient | Low | Normal | High | Excess |
|---|---|---|---|---|---|
| Nitrogen % (N) | 0.3[D]–0.8[D] | 0.9–1.2 | 1.3–1.5 | 1.6–1.8 | >2.0 (3.1[T]) |
| Phosphorus % (P) | <0.05 | 0.05–0.07 | 0.08–0.10 | 0.12–0.15 | >0.15 (0.45[T]) |
| Potassium % (K) | 0.06[D]–0.32[D] | 0.35–0.49 | 0.50–0.79 | 0.80–1.20 | >1.20 |
| Calcium % (Ca) | <0.3 | 0.3–0.4 | 0.5–0.8 | 0.9–1.1 | >1.1 |
| Magnesium % (Mg) | <0.05 (0.01[D]–0.04[D]) | 0.05–0.07 | 0.08–0.12 | 0.13–0.20 | >0.20 |
| Sulphur % (S) | <0.10 | | 0.18–0.25 | >0.25 | |
| Sodium % (Na) | | | 0.01–0.10 | 0.2–0.3 | >0.4 |
| Chloride % (Cl) | | | 0.01–0.20 | 0.3–0.6 | 0.7–1.5 |
| Copper[1] ppm (Cu) | <3 | 3–4 | 5–12 | 20–70 | |
| Zinc[1] ppm (Zn) | <9 | 9–14 | 15–50 | >50 | |
| Manganese[1] ppm (Mn) | <20 | 20–90 | 100–1500 | 1600–3000 | 3600[T]–5500[T] |
| Iron[2] ppm (Fe) | | | 25–200 | | |
| Boron ppm (B) | 8–12 | 13–19 | 20–50 | 60–80 | |

[1] Values for copper, zinc or manganese in leaves sprayed with fungicides or nutrient sprays containing trace elements cannot give a reliable guide to nutritional status even when leaves are washed.
[2] Leaf analysis is not a reliable guide to iron deficiency.
[D] Level found in field specimens showing deficiency symptoms.
[T] Level found in field specimens showing toxicity symptoms.
Main information sources and further reading: 1, 9, 10, 12

# Mango
(*Mangifera indica*)

**Sampling time** Early spring i.e. August–September (Northern Hemisphere January–February)

**Plant part and growth stage** Most recently mature leaf just prior to flowering

| Nutrient | Deficient | Low | Normal | High | Excess |
|---|---|---|---|---|---|
| Nitrogen % (N) | <0.7 | 0.7–0.9 | 1.0–1.5 | >1.5 | |
| Phosphorus % (P) | <0.04 | 0.05–0.07 | 0.08–0.20 | >0.25 | |
| Potassium % (K) | <0.25 | 0.3–0.4 | 0.5–0.9 | >1.2 | |
| Calcium % (Ca) | <0.4 | 0.4–0.9 | 1.0–3.5 | >5.0 | |
| Magnesium % (Mg) | <0.10 | 0.15–0.19 | 0.20–0.80 | >0.80 | |
| Sodium % (Na) | | | <0.20 | 0.30 | >0.40 |
| Chloride % (Cl) | | | <0.30 | 0.4–0.5 | >0.5 |
| Copper[1] ppm (Cu) | | 3–5 | 6–50 | >600 | |
| Zinc[1] ppm (Zn) | <15 | 15–18 | 20–150 | >250 | |
| Manganese[1] ppm (Mn) | | 25 | 30–500 | >800 | |
| Boron ppm (B) | | 20–24 | 25–150 | 200–300 | >300 |

[1] Values for copper, zinc or manganese in leaves sprayed with fungicides or nutrient sprays containing trace elements cannot give a reliable guide to nutritional status even when leaves are washed.

Main information sources and further reading: 1, 6, 9, 12

# Passionfruit
(*Passiflora edulis*)

**Sampling time**  During July–August for tropical and subtropical localities (Northern Hemisphere January–February)

**Plant part and growth stage**  Youngest fully expanded leaves on well developed actively growing laterals

| Nutrient | Deficient | Low | Normal | High | Excess |
|---|---|---|---|---|---|
| Nitrogen % (N) | <4.0 (1.2$^D$–2.9$^D$) | 4.0 | 4.5–5.5 | 6.0 | 6.8–7.5 |
| Phosphorus % (P) | <0.17 (0.09$^D$–0.13$^D$) | 0.17–0.20 | 0.25–0.35 | 0.4–0.6 | >0.7 |
| Potassium % (K) | <1.5 (0.3$^D$–1.2$^D$) | 1.5–1.8 | 2.0–2.9 | 3.0–3.6 | |
| Calcium % (Ca) | | 0.1–0.3 | 0.4 | 0.5–1.5 | >1.5[1] |
| Magnesium % (Mg) | <0.17 (0.04$^D$–0.15$^D$) | 0.17–0.20 | 0.25–0.50 | 0.55–0.70 | 0.8–0.9 |
| Sodium % (Na) | | | 0–0.15 | 0.20 | 0.3–0.6 |
| Chloride % (Cl) | | | 0–1.2 | 1.5–1.7 | >1.7 (1.9$^T$–4.0$^T$) |
| Copper ppm (Cu) | | 2–4 | 5–20 | 100–600[2] | |
| Zinc ppm (Zn) | <17 (12$^D$–16$^D$) | 17–20 | 25–100 | 150–290[2] | |
| Manganese ppm (Mn) | <20 (8$^D$–19$^D$) | 20–24 | 25–350 | 450–700[2] | 1400–3000$^T$ |
| Boron ppm (B) | | | 25–60 | | |

[1] If calcium values are above 1.5% check that the sampled leaves are not older than specified above.

[2] Many samples falling in this range are from healthy vines which have been sprayed with manganese, copper or zinc based fungicides.

Main information sources and further reading: 7, 12

# Pawpaw
### (*Carica papaya*)

**Sampling time**  Spring

**Plant part and growth stage**  Petioles from youngest fully expanded leaves subtending the most recently opened flowers

| Nutrient | Deficient | Low | Normal | High | Excess |
|---|---|---|---|---|---|
| Nitrogen % (N) | | 0.8–1.0 | 1.3–2.5 | >2.5 | |
| Phosphorus % (P) | | <0.2 | 0.2–0.4 | >0.4 | |
| Potassium % (K) | | 2.8 | 3.0–6.0 | >6.0 | |
| Calcium % (Ca) | | <1.0 | 1.0–3.0 | >3.0 | |
| Magnesium % (Mg) | | | 0.4–1.5 | | |
| Sodium % (Na) | | | <0.20 | | |
| Chloride % (Cl) | | | <4.0 | | |
| Copper[1] ppm (Cu) | | <4 | 4–10 | | |
| Zinc[1] ppm (Zn) | | | 15–40 | | |
| Manganese[1] ppm (Mn) | | 10–19 | 20–150 | | |
| Iron[2] ppm (Fe) | | | 20–80 | | |
| Boron ppm (B) | <16 (5$^D$–15$^D$) | 16–18 | 20–50 | >50 | 120$^T$ |

[1] Values for copper, zinc or manganese in leaves sprayed with fungicides or nutrient sprays containing trace elements cannot give a reliable guide to nutritional status even when leaves are washed.

[2] Leaf analysis is not a reliable guide to iron deficiency.

$^D$ Level found in field specimens showing deficiency symptoms.

$^T$ Level found in field specimens showing toxicity symptoms.

Main information sources and further reading: 8

# Persimmon
(*Diospyros* spp.)

**Sampling time** Early autumn i.e. March (Northern Hemisphere September)

**Plant part and growth stage** Youngest mature leaves from non-fruiting shoots

| Nutrient | Deficient | Low | Normal | High | Excess |
|---|---|---|---|---|---|
| Nitrogen % (N) | <1.0 | 1.1–1.4 | 1.6–2.8 | 2.9–3.1 | |
| Phosphorus % (P) | 0.05 | 0.06–0.09 | 0.10–0.26 | 0.27–0.31 | |
| Potassium % (K) | <0.8 (0.51$^D$) | 1.1–1.3 | 1.5–3.8 | >3.8 | |
| Calcium % (Ca) | <0.3 | 0.4–0.8 | 1.0–3.1 | | |
| Magnesium % (Mg) | <0.13 | 0.13–0.14 | 0.17–0.60 | 0.80 | >1.24$^I$ |
| Sodium % (Na) | | | 0.01–0.12 | 0.24$^I$ | >0.4 |
| Chloride % (Cl) | | | 0.1–0.6 | 1.2 | >1.5 (2.5$^T$) |
| Copper[1] ppm (Cu) | <1.5 | 2 | 3–37 | 250–380 | |
| Zinc[1] ppm (Zn) | <4 | 4 | 5–40 | | |
| Manganese[1] ppm (Mn) | <27 (7$^D$–15$^D$) | 27–30 | 40–500 | >600[2] | 1500–2400 |
| Boron ppm (B) | | <40 | 40–100 | | >390 |

[1] Values for copper, zinc or manganese in leaves sprayed with fungicides or nutrient sprays containing trace elements cannot give a reliable guide to nutritional status even when leaves are washed.

[2] Clark and Kajiura report 'green blotch' disorder of fruit where leaf calcium was low and manganese 600–2000 ppm, but no leaf symptoms occurred until leaf manganese exceeded 6000 ppm. (Clark, C.J. and Kajuira, I. *Growing To-day*, April 1986, pp 47–50.)

[I] Imbalance, from crop showing potassium deficiency symptoms.

[D] Level found in field specimens showing deficiency symptoms.

[T] Level found in field specimens showing toxicity symptoms.

Main information sources and further reading: 3, 9, 12

# Pineapple
*(Ananas comosus)*

**Sampling time**  During vegetative growth before flowering

**Plant part and growth stage**  Most fully matured 'D' leaf. Separate leaves into the middle third 'M' section, for nitrogen and chloride determination and a basal white tissue section 'B' for the remaining elements

| Nutrient | | Deficient | Low | Normal | High | Excess |
|---|---|---|---|---|---|---|
| Nitrogen % (N) | M | <1.5 | | 1.5–2.5 | | |
| Chloride % (Cl) | M | | | 0.2–0.8 | | |
| Phosphorus % (P) | B | <0.13 | 0.13 | 0.14–0.35 | | |
| Potassium % (K) | B | <2.8 | 2.9–4.2 | 4.3–6.4 | >6.4 | |
| Calcium % (Ca) | B | <0.04 | 0.05–0.21 | 0.22–0.40 | >0.4 | |
| Magnesium % (Mg) | B | <0.13 | 0.14–0.40 | 0.41–0.57 | >0.57 | |
| Sodium % (Na) | B | | | <0.02 | | |
| Copper[1] ppm (Cu) | B | | | 10–50 | | |
| Zinc[1] ppm (Zn) | B | | <20 | >20 | | |
| Manganese[1] ppm (Mn) | B | | <80 | 150–400 | | >1200 |
| Iron[2] ppm (Fe) | B | | | 80–200 | | |
| Boron ppm (B) | B | | <30 | 30–80 | | |

[1] Values for copper, zinc or manganese in leaves sprayed with fungicides or nutrient sprays containing trace elements cannot give a reliable guide to nutritional status even when leaves are washed.

[2] Leaf analysis is not a reliable guide to iron deficiency.

The standards and procedures used for pineapples are largely those of Consolidated Fertilisers Ltd Queensland (Reference 1).

M = middle third of D leaf; B = basal white tissue of D leaf

# References

1. Anon. (1989). Plant tissue analysis service interpretation. *Agricultural Manual Vol. III.* Incitec Ltd, Brisbane.

2. Benton Jones, J., Wolf, B. and Mills, H.A. (1991). *Plant Analysis Handbook.* Micro–Macro Publishing Inc., Georgia, USA.

3. Clark, C.J. and Barrett, L.J. (1985). Leaf analysis of persimmons. *Growing Today.* **Aug.**, pp. 30–31.

4. Cresswell, G.C. (1989). Development of a leaf sampling technique and leaf standards for kiwifruit in New South Wales. *Aust. J. Exp. Agric.* **29**, 411–17.

5. Goodall, G.E., Embleton, T.W. and Platt, R.G. (1965). Avocado fertilisation. California Agric. Expt. Stat. Leaflet 24.

6. Martin-Prevel, P., Gagnard, J. and Gautier, P. (1984). Plant analysis as a guide to the nutrient requirements of temperate and tropical crops. Lavoisier Publishing Inc., New York and Paris.

7. Menzel, C.M., Winks, C.W. and Simpson, D.R. (1989). Passionfruit in Queensland, 3. *Qld Agric.* J. **May–June**, pp.155-64.

8. O'Hare, P. (1993). *Growing Pawpaws in South Queensland.* Queensland Department of Primary Industries.

9. Reuter, D.J. and Robinson, J.B. (1986). Plant Analysis – An Interpretation Manual. Inkata Press, Melbourne.

10. Stephenson, R.A. and Cull, B.W. (1986). Standard leaf nutrient levels for bearing macadamia trees in south-east Queensland. *Scientia Horticulturae* **30**, pp.73-82.

11. Turner, D.W. (1985). Fertilising bananas: leaf analysis as a guide. Agfact H6.3.5 Department of Agriculture, N.S.W.

12. Weir, R.G. and Cresswell, G.C. (Unpublished). Diagnostic and research leaf analyses records. Chemistry Branch NSW Agriculture 1958–1988.